WITHDRAWN
UTSA LIBRARIES

FARMING
AND FOOD SUPPLY

FARMING AND FOOD SUPPLY

THE INTERDEPENDENCE OF COUNTRYSIDE AND TOWN

by

SIR JOSEPH HUTCHINSON

C.M.G., SC.D., F.R.S.

Drapers' Professor of Agriculture Emeritus,
University of Cambridge

CAMBRIDGE

AT THE UNIVERSITY PRESS

1972

Published by the Syndics of the Cambridge University Press
Bentley House, 200 Euston Road, London NW1 2DB
American Branch: 32 East 57th Street, New York, N.Y.10022

© Cambridge University Press 1972

Library of Congress Catalogue Card Number: 79–184140

ISBN: 0 521 08475 X

Printed in Great Britain
at the University Printing House, Cambridge
(Brooke Crutchley, University Printer)

CONTENTS

ACKNOWLEDGEMENTS

This book had its origin in lectures on comparative agriculture given over a period of twelve years when I was Drapers' Professor of Agriculture in Cambridge. The stimulus to study and to teach in this field came from Professor Sir Frank Engledow FRS, of whom I was once a pupil, and whom I succeeded in Cambridge. I inherited the lecture course from him, and his notes formed the basis on which I developed my own ideas. I cannot adequately thank him.

In the winter of 1969–70 I held a Royal Society Leverhulme Visiting Professorship at the Indian Agricultural Research Institute in New Delhi, and I undertook to give a course of lectures there on the same subject. My notes as I had by then developed them proved surprisingly inadequate as a basis for lectures on the same topics in New Delhi as I had discussed in Cambridge. If in this book I have been able to achieve some breadth of view, it is largely because of my good fortune in lecturing on the same topics in two continents. I am grateful to Dr M. S. Swaminathan, the Director, and his staff for the welcome they gave me and the facilities they put at my disposal.

My obligations in respect of information supplied and criticism offered are too numerous to set out in full, but I must acknowledge my debt to F. Hardy, one of the wisest of all students of tropical agriculture, from whom many others beside myself first learnt the principles of tropical crop ecology.

Finally, I owe a great deal to F. A. Buttress, Librarian in the School of Applied Biology in Cambridge, for his generous and unfailing help in all library matters, and in particular in locating little known but important references.

J.B.H.

PREFACE

Through agriculture, human communities have established a situation in which they are relieved of the necessity to spend the greater part of their time in pursuit of the necessities of existence. One man is able to feed a number of others beside himself, and thereby to liberate the others to do other things, in part on his behalf. So agricultural specialists feed craftsmen and artisans, priests and civil servants, who have left the land and developed urban communities. A rising standard of living has depended historically on the ability of agriculture progressively to release men to these other activities, and with its own reduced human resources to feed the increased urban population. This is a process that has gone on ever since agriculture began, and which continues today. Even in Britain, in which less than four per cent of the population is engaged in agriculture, the industry is still releasing workers at a rate of five per cent of its own work force per annum.

Impressed by this continuing reduction in numbers, some authorities have regarded agriculture as a declining industry. An industry that produces the basic essentials of existence, and does so with increasing efficiency with a steadily smaller demand for human effort, is not a declining industry. That it is a mis-understood industry is not to be denied, and it is the purpose of this essay to reduce this misunderstanding and to trace the achievements of agriculture in the promotion of human welfare.

The early stages in the process whereby agriculture changes and develops to support increasingly diversified communities can still be seen in many parts of the world. In Britain, and in only a few other places, it is possible to study the most recent stages of development that have led to spectacular increases in productivity. Increases in productivity of this order will be needed in many other parts of the world if the rising population is to be fed, and higher standards of living are to be achieved. Hope for the future is sustained by the success of India, Pakistan and some other Asian countries in raising their production and approaching a balance in their food budgets. It is worthy of note that the countries that have made good progress in this respect are those with soils good enough and robust enough, and climates sufficiently reliable, to have supported large populations over several millennia. Indeed it appears that with the ready availability of industrial fertilisers, the prospects of increasing food supplies are best on good lands with a good climate that are already in cultivation, rather than on the marginal lands that have not been brought into agricultural use. Hence it is appropriate in this study of agricultural development to take Britain, with a long history of improvement of intrinsically good land, as the standard of comparison.

My own experience has been in Britain, the West Indies, parts of Africa, and India, with short visits to other countries. For this reason the comparisons I have made are for the most part between Britain, Africa and India. I have endeavoured to write an essay in which are set out the views and convictions impressed upon me by my personal experience. I hope many of them will gain assent, and for the rest, if they generate debate on important issues, I shall be well content.

1971   J.B.H.

PART I: INTRODUCTION

CHAPTER I

THE HISTORY OF
AGRICULTURE

The origin and development of agriculture is of major significance to archaeology, to agricultural science, and to human history. It is an evolutionary phenomenon. It has given rise to new races, both of plants and of animals, at a rate of change that it is very difficult to match anywhere else in the plant or animal kingdoms. Under its influence, man has also undergone rapid social evolution, and though we have very limited means of assessment, there is little doubt that he has enjoyed substantial genetic change also during the period of agricultural activity.

Agriculture was not the first of man's great achievements. As a hunter and food gatherer he had colonised the world very widely before he became a farmer and he had attained a high standard of technology in making and using tools of stone, horn and bone, and doubtless also of more perishable materials of which we now no longer have evidence. The beginnings of agriculture have only recently been identified. The evidence available to us is archaeological, and consists of the refuse of human communities; bones, grinding stones, pottery on which imprints of grain may be found, and remains of stores of grain or dumps of damaged foodstuffs. There seems good reason to suppose that hunters first became herders, exercising increasing control over the animals they used to hunt until they had established flocks and herds that were dependent upon them for protection, and for the provision of, or for guidance to, food and water. Such a change leaves little evidence in the archaeological record. Whether the goat or sheep bones in a midden were those of a wild animal that was hunted, or a domestic animal that was slaughtered, is impossible to determine with certainty. Opinion can only be based upon frequency of occurrence, proportion of the whole sample that belongs to the putative domesticate, and so on. However, in the later stages the interdependence of the human community and the livestock population becomes evident beyond dispute. The association between man and the wolf/dog species was established by c. 9000 B.C. Goats and/or sheep were closely associated with man by 7000 B.C. at Jericho, and by 6600 B.C. at Jarmo.

'Our present knowledge of the earliest plant husbandry is derived from four sites in the Near East, of about 7500 to 6500 B.C., viz., Ali Kosh [Khusistan, Iran], Beidha [Palestine], Hacilar in west-central Anatolia [Turkey], and Jarmo in northern Iraq.' (Helbaek, 1966.) These sites were occupied by pre-pottery Neolithic people. A similar culture arose in Greece, and Jane Renfrew (1967) has reported on grain recovered from aceramic Neolithic sites there. In both areas the main cereals were Emmer wheat (tetraploid) and hulled two-row barley. The Near East is the area of distribution of the *Hordeum* and *Triticum* species ancestral to these early cultivated cereals, and in the pre-pottery Neolithic cultures of this area we have the very early stages of agricultural development, the adoption and fostering of wild grasses which developed under care and selection into our major cereals.

Though crop production was based on barley and wheat, a range of other plant remains indicates on the one hand the continued importance of food gathering (Helbaek and Renfrew report finds of acorns and pistachio), and on the other the beginnings of domestication of other crop plants (Helbaek and Renfrew report peas, beans, vetches, oats, and lentils). Moreover, Helbaek has deduced from spikelet morphology that the cereals included a range of types from near-wild forms with a brittle rachis to more advanced forms with the tough rachis characteristic of cultivated races.

Having identified a centre of origin of agricultural practice, one naturally enquires whether it was unique. It may well be argued that a much wider area than the Near East was involved in those beginnings. The Nile Valley, Ethiopia, and Iran and Central Asia, all had agriculture at an early date, and the Egyptian records in particular go back to the Neolithic in the Fayum. Nevertheless, the evidence from crop plants and domestic animals is strong that the agriculture of the Old World originated in a rather limited area in the Near East including upper Iraq, Palestine and Turkey. This is the area in which the wild relatives of wheat and barley are to be found, and also wild relatives of the cultivated legumes and the major species of domestic livestock: sheep, goats and cattle. Moreover, though cattle are not indigenous in Africa, all the oldest agricultural communities on the continent had them, and it must be presumed that they were brought in from the Middle East.

One other area in which agriculture started independently, and apparently within about 2000 years of the same date, was in lowland, semi-arid Mexico. MacNeish (1964) has excavated caves in the Tehuacan valley, and has reconstructed the beginnings of agriculture there about 7000 years ago. In collaboration with agricultural botanists, particularly Mangelsdorf, he was able to identify the wild ancestor of maize, and to trace the development of the cultivated crop plant from it. He also observed the evidence of the beginnings of cultivation of cucurbits, and of species of *Phaseolus* beans.

Both the barley and wheat system and the maize and beans system originated and developed in areas with a rather dry or even semi-arid ecology, where grasses were the natural dominants under an open tree cover. In tropical forest areas there are farming systems based on root crops, and it has been suggested that they may have arisen independently in several places, notably on the borders of the Amazon forests in South America, and in South East Asia. The history of agriculture in such regions is difficult to determine, since organic remains in these wet countries are lost from the archaeological record.

In both areas from which we have good evidence, agriculture developed on a sound nutritional base. In the Near East, calorie supplies from cereals were balanced with protein supplies from cereals and from meat. In Tehuacan, maize provided calories, and beans protein. In both, protein supplies were doubtless supplemented by hunting. The spread of agriculture did not, at least initially, involve any sacrifice of nutritional levels. Even in the early agricultural systems based on roots and tubers, with abundant calorie supplies and low protein, the diet was balanced in many places – in the Amazon and Orinoco valleys, on the Caribbean coasts and the West Indian islands, and in coastal South East Asia – by protein from fish.

Consider now the Old World. Once farming had begun, it

3

spread out from its Middle East centre. It reached Greece very early. It was established in Italy, Central Europe and Germany by 4000 B.C. and in France, Denmark and the British Isles by the latter part of the fourth millennium B.C. (Clark, 1965). New crops were established, including oats and rye, *Brassicas* and beet; and horses and pigs were domesticated as new needs arose and new opportunities offered. The process went on slowly. Iverson (1941) records that the only cereals grown in Denmark in the Late Stone Age were wheat and barley. Oats and millet (*Panicum miliaceum*) appeared first in the Bronze Age, and rye in the Iron Age.

In Western Europe extensive studies by botanists and archaeologists have elucidated the timing and nature of the arrival of agriculture. Farming could only be practised if land could be cleared of forest (Clark, 1945). Evidence of clearings is found about 3000 B.C. all over Western Europe, in a decline of tree pollen in pollen diagrams from lake deposits, often associated with a fine carbon layer. The decline of tree pollen is followed by an increase in the pollen of such species as *Plantago lanceolata* that are common weeds of arable fields. Iverson (1941) has shown that the Neolithic farmers practised a cut and burn shifting cultivation. The carbon layer was from the burn, and the change in the pollen diagram followed from the destruction of the forest, and, after cropping, the abandonment of the land and the shift to a new area. Neolithic farmers found, as shifting cultivators in Africa find today, that after a few seasons low soil fertility and rampant weed growth compelled the abandonment of the old clearing and the felling of a new area. The close correspondence between current practice in Africa and this reconstruction of Neolithic practice in Europe is of great significance. It establishes shifting cultivation as one of the oldest and most persistent systems of agriculture. More than this, if some Western European soils were originally such that only shifting cultivation could be practised, their improvement to the point where they support the current high productivity of Western European agriculture gives grounds for believing that at least the better African soils may be susceptible to similar fertility amendment.

Clark (1965) has remarked on the comparative tardiness of the spread of Neolithic culture south into the Nile valley – and thence into the African continent – compared with the early and vigorous invasion of Europe, north of the Mediterranean. This rather later agricultural colonisation gave rise to the establishment of crops and stock of European provenance on the Ethiopian highlands, and the domestication and improvement of the African range of crop plants, including *Sorghum* (dura in the Sudan; jowar in India), *Pennisetum* (bulrush, or pearl, millet; bajra in India), *Eleusine* (finger millet; ragi in India), *Dolichos lablab* (beans), *Vigna* (cow peas), and *Dioscorea* (yams). The consequences of this for agricultural botany were recognised by Vavilov, in his designation of Ethiopia as a secondary centre of diversity for many of the Middle Eastern crop plants. For the African crops on the other hand, Africa was recognised as the primary centre.

The south eastwards spread of agriculture brought farming peoples into the Indian sub-continent. Archaeological data are limited at present and further excavation is expected to lead to substantial increases in knowledge. The information now available is summarised by Allchin (1969) and by Vishnu-Mittre (1968). Excavations in Baluchistan have revealed an early agriculture, based on wheat, dated as late fourth millennium B.C. In the north west of the Indian sub-continent (West Pakistan, Kashmir, Punjab, Sind and Rajasthan), the people of the Harappan civilisation (2300–1750 B.C.) practised farming based on wheat, barley and peas of Middle Eastern origin, and sesamum and cotton of local domestication. The extension of farming beyond the Indus valley depended substantially on the addition of new crops to those brought in from the Middle East. Rice first appears in Gujarat in Harappan times. Vishnu-Mittre records *Sorghum* (jowar) from Sind, and *Pennisetum typhoides* (bajra) from Gujarat, also in Harappan times. These cereals are of African origin. They must have been introduced, and there must have been a farming community beyond the Nile valley early enough to have domesticated them and to have passed them on to the Harappan culture.

An agriculture equipped with such a diversity of crop plants, from the Middle East and Africa as well as indigenous, can hardly have been the earliest in the region. Moreover the fragment of cotton textile found at Mohenjo-Daro and examined by Gulati and Turner (1928) was the product of a sophisticated textile craft. Evidently, farming cultures earlier than the Harappan are still to be discovered.

It was rice that made possible the colonisation of the wet lands, and the African cereals, *Sorghum* (jowar), *Pennisetum* (bajra) and *Eleusine* (ragi) that led to the establishment of farming on the rainfed lands of the peninsula. In the post-Harappan (Chalcolithic) age (1750–1000 B.C.) rice spread widely. The third African cereal, *Eleusine* (ragi), appeared, together with lentils and linseed from the Middle East and the indigenous species of *Phaseolus* and *Dolichos* (pulses). The sites from which agricultural remains of this period have been recovered extend as far south in the peninsula as Mysore.

In livestock also, India added substantially to the list of domesticates. Sheep, goats and cattle may have come in with the earliest farmers, but the zebu was domesticated in India from native wild stock, as was also the buffalo. And by 2000 B.C. India had domesticated the jungle fowl, to give the ancestors of the modern hen.

Spread in the New World was necessarily north and south. It gave rise quite early to the South American cultures on the west coast and on the Andes. Their importance lies mainly in the substantial additions to the list of crop plants that were made in South America, the tomato, the Lima bean, the groundnut and the New World cottons, and also the high altitude group of crops including *Chenopodium*, potato, *Oxalis*, *Tropaeolum* and ulluca. From South America also came the guinea pig and the domesticated cameloids. It is worth noting that America north of the original centre of agriculture has added nothing except perhaps the turkey to the repertoire of crops and stock.

The practice of agriculture released the human population from a very strict limitation on its numbers. In the New World, MacNeish (1962, 1964) has estimated the population in the Tehuacan valley at successive stages in its cultural development. As hunting and food gathering bands, there was subsistence for only a very few, widely scattered family groups. They occupied large territories over which they roamed according to season and to consequent supplies of food and water. With the beginnings of farming, greater production, storage of food, and the elimination of the need for nomadic wandering, made population increase possible. The small family bands became small communities. As they expanded they met other like communities, and MacNeish gives reasons for suggesting that different bands having domesticated different plants, the range of cultivated cucurbits and legumes arose from the exchange of their separate domestications between groups. Irrigation early added yet another major resource to the community, and the development of crafts became possible with the enhanced food supply. Over 8000 years MacNeish estimates that the carrying capacity of the Tehuacan valley increased about 5000 fold. Thus agriculture made possible the emergence of large human communities in place of the scattered hunting bands of the palaeolithic. The production of more food with less effort released some people to specialise in crafts and services, and gave rise for the first time to groups that did not grow their own food. These were the beginnings of towns.

The spread of agriculture over the world led to the establishment of farming communities in very diverse circumstances and the subsequent history of agriculture is in fact the history of local adaptation and of progress stimulated or inhibited according to local circumstances. Nevertheless there is throughout a common thread, which is the interaction between farming on the one hand and crafts, industries and services on the other. The demand of urban communities for food and for industrial crops has been the great stimulus to agricultural innovation, and agriculture has been equipped for high production with the products of urban crafts and industries. The relationship between rural and urban occupations is one of complementation and mutual stimulus.

One of the great episodes in British agricultural history was the agricultural revolution that is called the New Husbandry. On the farms of eastern England, consolidated by the enclosures, a group of far-sighted landlords and farmers initiated in the latter part of the 18th century a new system of crop husbandry, linked with a much intensified system of stock keeping. The intensification of the crop rotation and the improvement in both feeding and breeding of livestock were but a part of the New Husbandry. An essential feature of the system was the careful conservation of the elements of fertility by folding sheep on the land and by wintering cattle in yards and distributing the manure on the fields. The New Husbandry was developed under the stimulus of the Industrial Revolution, with its growing urban population, increasing in wealth, and demanding larger and better supplies of food. Moreover, as farming and industry developed, the manufacture of implements and machinery contributed industrial resources to the advancement of farming.

Recent agricultural history can only be studied in terms of the relation between farming and the urban markets which it supplies. In the advanced, high productivity agriculture of Western Europe and North America is to be seen the integration of productive farming with productive industry, with the greatest development of the use of industrial inputs to gain high farm output. By contrast, the low yield agriculture of most of Asia is geared to static and tradition-bound urban communities which contribute only the traditional inputs of a static and exploitive agriculture. Only where modern industrialisation has gone on, and where modern medicine has cut down death rates and given rise to a population explosion, has there arisen the stimulus of changing and increasing urban demand. There, after an initial delay that raised serious problems in food supply, agriculture is responding with increased productivity. In great areas of the world, however, there are still communities that are only one step ahead of self sufficient subsistence. These are countries with poor land, small populations, and only the beginnings of urban development. They have advanced initially by producing crops that could be sold in the markets of the world's industrial countries, and they have had in exchange the infrastructure of development: law and order, education, transport systems, and a limited supply of consumer goods.

In the advanced countries all the major establishment costs of agriculture have been met and written off. The land has been cleared and levelled, water control undertaken by drainage or by irrigation, and a sound system of soil management set up. It has been calculated that in Britain, if bare land were taken and capitalised with roads and buildings, at current farm incomes the investment would only pay if the land cost nothing. This amortised capital is a great advantage. The emerging countries have only gone part way along the same line. In opening new land, and in developing old land to higher standards of productivity, this basic capital has to be found, put in, and written off in due course.

The beginning of modern agriculture in Britain was the establishment of conservation by the closing of the fertility cycle, and returning to the land those elements of fertility that are removed in cropping and may be returned as human and animal wastes. The first step in agricultural advance in any developing country is the establishment of the principle of conservation. This includes conservation of soil, of water, and of nutrients. Fertility levels are stabilised in various ways. In shifting cultivation with a sparse population, fertility decline under cultivation is balanced by regeneration under bush or forest, provided that the period of regeneration is long enough. Under ancient farming systems with a dense population, the tendency is for land to be degraded to a basic fertility level, where losses are balanced by gains from casual return of wastes to the land, from the decay of parent rocks, and from biological fixation of nitrogen. Under such circumstances there may arise forms of fertility transfer, as in the concentration of fertility round Indian villages. This is what happens on the better and more stable lands. On less stable land, there may be no end to degradation, with 'bad lands', gullied and almost impassable, or rock surfaces left behind and the human population gone. Stopping this kind of accelerated soil deterioration is not unduly difficult. The next consideration is to hold the water on the land long enough to allow adequate percolation into the soil. Water conservation is a major contribution to stability. It may indeed alter the situation out of all knowledge, as for instance in northern Nigeria where water control made worthwhile pest control and the application of fertilisers and manure (Lawes, 1968).

Conservation of the elements of fertility follows. Composting of all available organic wastes was advocated by Howard at Indore in India in the early 1930s, but it was never adopted in Indian farming practice. In current thinking there is a strong tendency to pass over the conservation stage in the amelioration of tropical soils and go straight to the stage of improvement by fertilisers. Improvement by chemical fertilisers pre-supposes soil in good structural condition to begin with. In drawing on temperate region experience, insufficient allowance has been made in the assessment of the advances gained from fertilisers,

9

for the sound state of the soils to which they have been applied. The change in Norfolk farming, for instance, in the 1920s and 1930s (see Rayns, 1961) was mounted on the soil conditions built up by more than a century of conservation of the elements of fertility under the principles of the 'New Husbandry' established by the Norfolk improvers. In the developing countries this conservation period has not been experienced, and in attempting to get from an exploitive to an improving agricultural system in one stride, the conservation of organic matter must not be overlooked.

Fertility improvement depends on the maintenance of a balance both between fertiliser elements, and between fertilisers and crop varieties. When in 1916 Biffen at Cambridge released Yeoman, the first of the high yielding wheats, its success was limited by the limited amounts of land of high enough fertility to exploit its merits. Now that high fertility levels have been established in Britain, their exploitation depends on the strains of crop plants of high yield potential, that have been bred in recent years. History repeats itself. In India the new wheats, like Biffen's Yeoman, only yield as they should on highly fertile soil and with high fertiliser application. The historical record is clear. High fertility levels and high yield potential must go together, and the plant breeder and the agronomist have always been partners in the improvement of productivity.

In the long history of advancing productivity in British farming, the improvement of livestock has gone on in parallel with the improvement of crops. Stock and crops were integrated in a single farming system, partly because livestock – first oxen and then horses – were the source of power on the farm, and partly because the system involved the use of grazing, both of waste land and of sown crops, for the production of meat or dairy produce. The integration served also the purpose of fertility conservation, since much of the fertility taken from the soil was returned as manure.

In modern times the replacement of horses by tractors as a source of power, and the advocacy of specialisation on economic grounds have led to the abandonment of mixed farming on many farms. The simplicity of a system in which a farm is organised for crops only, and the advantages to a farmer seeking to equip his farm with a limited outlay of capital, are attractive.

Nevertheless, in a farming system geared to the provision of food for an affluent and sophisticated urban society, a very large part of the crop product must be fed to stock, since livestock products form such an important part of the diet. So stockless farms are matched by intensive pig and poultry enterprises, by barley beef, beef lots, and large and intensive dairy units. What used to be the systematic return of the elements of fertility to the land by way of the manure cart has become a major problem in effluent disposal. Moreover on the stockless farms, problems of the maintenance of good soil structure are increasing. Fertiliser applications are disproportionately high where fertility conservation is not practised, and under near monoculture farming, inputs of pesticides and herbicides are at a level that is expensive, and damaging to the wider environment.

This is a historical survey, and it would be inappropriate to speculate on future trends. It is proper to remark, however, that the tendency to model agricultural development in developing countries on the most recent pattern of the advanced countries of Western Europe and America is unfortunate. The great advances in western agriculture were made by those who devised the system of fertility conservation known in Britain as the New Husbandry, and the high production of modern British agriculture still is founded on the excellence of the soils established thereby.

PART II: THE COMPONENTS
OF AGRICULTURE

CHAPTER 2

CLIMATE AND SOIL

Climate and soil are the two major natural factors dominating the practice of agriculture. The areas in which farming began are not those in which it was subsequently most successfully developed. The semi-arid grasslands exist because of the limitations of rainfall, both in amount and duration. Areas of greater rainfall and longer rainy season are naturally clothed in forest, and when man acquired the ability to destroy natural vegetation he began to encroach upon the forest and to exploit the natural climatic advantages of the forest lands of the temperate regions. Thus much of the spread of agriculture has involved the destruction of forest and the imposition of a grassland ecology in its place. In Britain, forest clearance has gone on throughout agricultural history. The changes thereby brought about have often been deplored by historians and by naturalists, and have from time to time been halted or even temporarily reversed. William the Conqueror reversed the process in the New Forest. There was a famous 18th century admiral who always filled his pockets with acorns when visiting the country, so that he could sow the beginnings of good ship timber for generations to come. There are loud complaints at the destruction of hedgerows and hedgerow timber, many of which are in fact a comparatively recent addition to the British countryside.

Forest destruction has gone on far beyond Britain and Western Europe. Even where cereal farming is not practised, tree and bush cultivation still require the removal of the forest. Forest clearance in the West Indies for agriculture based on sugar-cane and bananas, and in Central and South America for coffee, are recent examples. Progressive degeneration of forest under the onslaught of agriculture can be seen in Ghana in the cacao belt. Indeed throughout the world the forest is in retreat as man seeks to meet the needs of his vast and increasing populations.

Clearance alone does not make an agricultural soil. Clearance in fact creates problems in both water relations and soil management which must be solved if the productivity of the land is to be ensured. Natural vegetation has developed under the selective forces of the climate, and under the influence of the soil as a reservoir of water and of nutrients. Climate is beyond man's control, and his agriculture must be planned within the limits set by it. Soil, on the other hand, can be amended, though it cannot be fundamentally changed.

The limitations imposed on agriculture by climate will be considered first, and within these limitations, the domestication of soils will then be discussed. The two attributes of climate that dominate vegetation, both natural and agricultural, are temperature and rainfall. Temperature governs mainly the length of the potential growing season. The tropics are warm enough for growth the year round. In higher latitudes a cold season limits growth progressively, until in Arctic and Antarctic regions no vegetation exists. The relation between temperature and latitude is changed by altitude, lands at high altitudes being cooler than lands at sea level at the same latitude.

Within these major temperature limitations, agriculture is confined to areas in which the water supply is not only sufficient for crop growth, but sufficiently reliable to ensure successful cropping in all, or very nearly all, years. This means areas either with a reliable as well as an adequate seasonal rainfall, or with a reliable source of water for irrigation. Natural vegetation is not so restricted by reliability, and desert vegetation, for instance, is so organised as to grow when there is moisture available, and to remain dormant when it is not. It carries over the drought period as dormant seeds, thorny scrub, or underground storage organs. Moreover, where rain is more adequate, natural vegetation is selected to fit the pattern of the rainy season in a way that can rarely be matched with cultivated crops.

Considerations of water use in agriculture fall under two heads: the amount, distribution and reliability of rainfall on the one hand, and the need of crops for water on the other. Comprehensive and reliable information on rainfall is vital to the planning of agricultural development, whether it be the extension of farming into uncultivated land in Australia or the intensification of agriculture by supplementary irrigation in Britain. The importance of rainfall reliability is well illustrated by the ill-fated Kongwa enterprise of the groundnut scheme in Tanganyika. The best available rainfall data indicated a rainy season of about seven months, with a mean annual total of 600 mm. This would be adequate for a groundnut crop, but the variability of the rainfall is such that in no month of the year can as much as 25 mm of rain be confidently expected, while in any of the 4 months January to April there may be more than 200 mm. In fact there is a risk of both a drought and a flood in the same season. Moreover, with the high frequency of very big storms the distribution of monthly rainfall is skew and the most likely rainfall total for a year is not 600 mm but 500 mm which is very much less adequate.

Various statistical techniques have been used for the assessment of rainfall reliability. Glover and Robinson (1953) have concerned themselves with the chances of obtaining a satisfactory seasonal total. Skewness is small with totals, so with a modest amount of computation they were able to prepare maps of East Africa giving the expectation of obtaining (a) 20 inches, and (b) 30 inches of rain per annum. Similar maps for the United Kingdom are given by Gregory (1957). A more precise assessment of reliability is necessary in studying crop potentialities in detail. One was devised by Manning (1950 and 1955) on the basis of monthly data. In view of the skewness of distribution of monthly rainfall, this involved a rather more elaborate statistical treatment. After transforming to remove the effect of skewness, Manning was able to calculate confidence limits for monthly rainfall, and to prepare rainfall diagrams, showing upper and lower limits through the season, at any predetermined confidence limits. He took two criteria: 9:1 as giving a risk of transgressing the lower limit once in 20 years, and 1:1, with a risk of transgressing the lower limit once in four years. The former gives a fair assessment of the risk a peasant farmer cannot afford to exceed, and the latter is more nearly what an advanced and capitalised agriculture could face. Studies of the incidence and reliability of rainfall of this kind make possible first a more efficient use of rain, by matching the timing of the crop to the pattern of the rain, and second an assessment of the risk of crop failure, and hence a greatly increased confidence in the planning of development.

The establishment of the principles on which to determine the demand for water by vegetation is one of the great recent advances in agricultural science, and it has had a profound effect on our understanding of the factors governing the agricultural use of land. In the 1920s, American workers began to study the needs of plants for water at the desert laboratory at Tucson, Arizona. They were concerned with the evaporative power of the air, and Livingston devised a system of measurements of evaporation from porous pots, some of which were white and reflected light, and some black which absorbed most of it. Though much information was obtained, no general principles emerged, and the technique fell into disuse because of difficulties such as changes in porosity due to dust deposit, growth of algae and so on. Hydrologists concerned with water conservation in reservoirs developed the use of open pans from which evaporation was measured, and a standard 48 inch diameter pan has come into general use. The water losses it records vary considerably according to the siting and mounting of the pan. The relevance of the data to evaporation from vegetation was until recently entirely conjectural.

The great advance in recent years came from Penman's (1948) concept of evaporation as a physical phenomenon, and therefore calculable from physical data. The evaporation of water takes a definite and known amount of energy per unit weight. Penman devised formulae whereby the maximum amount of energy available for the evaporation of water could be calculated from standard meteorological data. He checked his findings against data for evaporation loss from a close-cut grass turf, freely supplied with water. He was able to show that there was good agreement between predicted requirement and actual use.

Some of the conclusions that follow from Penman's interpretation of water use by plants, agriculturists found it difficult to accept. That there was an upper limit to the amount of water plants could use was reasonable. But that the upper limit was the same in any given land-and-climate circumstances, regardless of whether the vegetation was marshland, grassland, cereals or forest, was very difficult to concede. In Uganda, for instance, *Eucalyptus* is planted regularly and successfully, as a means of drying up swamps. In South Africa tree planting has resulted in the disappearance of what were perennial streams in the grassland, before the trees were planted. It was difficult to reconcile the vast amounts of water needed to grow a crop of cotton, with the meagre supplies on which acacia thorn scrub survives in the Sudan desert, if it is all a matter of the energy input to the area.

Penman's concept is now generally accepted, and acceptance has come through recognising that his calculations estimate maximum water use when water is freely available. A 10 m *Eucalyptus* cannot lose by evaporation any more water than the energy available will evaporate, and a 2 m swamp grass may evaporate the same amount. But if the root range is in any way proportional to the above ground size of the plants, the *Eucalyptus* will be evaporating water in periods where the grass does not because all the water within root range has been used. Likewise, the desert acacias survive on little water, but to grow a crop on the same land, a copious and regular supply of water must be provided. The acacias would use more water if it was available, but a crop, though it might survive, would not yield if restricted to what the acacias get.

Penman's calculations provide estimates of the energy per unit area available for the evaporation of water. Where the area is covered with leafy vegetation, water is lost by transpiration. Where the ground is bare, it is lost by direct evaporation from the soil. The sum of these factors is termed evapotranspiration. The potential evapotranspiration is only reached where water is freely available over the whole area. Thus, on a desert, the surface soil is dry and direct evaporation loss is negligible. If there is water available within root range, transpiration will continue from such vegetation as is established and can reach it. In arable agriculture, water loss from a seedling crop is far below potential evapotranspiration because the leaf area of seedlings is small, and the surface soil dries rapidly and direct evaporation is then much reduced. Evapotranspiration per unit area rises rapidly as leaf area expands until virtually all the incident energy is intercepted and evapotranspiration reaches the limit set by the energy available. Transpiration goes on only from living tissues, so as senescence sets in, the rate falls, and water use declines.

Discrepancies between calculated and actual water use have also arisen from factors which alter the energy supply as against the estimates from meteorological data. In temperate regions a common cause is the surface 'roughness' of a crop – e.g. Brussels sprouts – as against the smooth surface of a lawn or grass field. The resultant turbulence alters the energy relationships and energy may be released for the evaporation of water by the cooling of the turbulent air. In dry tropical climates, a fetch of wind over a bare hot desert provides considerable extra energy – the 'advective energy' of a hot wind – over that estimated from the meteorological data.

Water deficit can now be accurately measured, and work is going on actively on the measurement of the effects of drought, both in general and at particular periods of crop growth. The agricultural effectiveness of rainfall may be defined as the chance of an adequate supply of water being available to meet crop needs throughout the growing season. And crop needs will not just be equivalent to the Penman calculation of evapotranspiration, but to this adjusted for the area of active transpiring surface available. With the recognition of the physical basis of water use it has become possible to estimate from meteorological data the potential water use by vegetation month by month for any climate. These figures can be compared with the monthly rainfall expectation, and a rough 'demand and supply' relationship arrived at. In arid areas, such a figure gives, first, an estimate of the potential growing season and, second, a figure for the amount of water required to provide for unrestricted crop growth if irrigation is to be supplied. Of equal importance is the calculation of the 'demand and supply' relationship in heavy rainfall areas, both in the wet tropics and in the wet zones of the temperate regions. The disabilities of agriculture in areas of insufficient or unreliable water supply are well known. The deleterious effects of an excess of water over crop requirement are not so fully appreciated. Water in excess of that used by vegetation drains away, either by run-off, with all the hazards of erosion, or by percolation, with consequent leaching of soluble nutrients.

Where there is a substantial excess of water supply over vegetation needs, the establishment of agriculture has been difficult, and often completely unsuccessful. Probably the best documented attempt to amend the fertility status and establish an improved agricultural system on land with a rainfall excess is the work initiated by Stapledon on the improvement of grass land on the hills of Wales. Under the permanent grass cover, run-off does not lead to erosion but the rate of loss of nutrients by leaching is high. Pasture production can be substantially improved by fertiliser use, but the improvement is ephemeral, and can only be maintained by regular treatment to make good the losses due to rapid leaching. This is done at a cost that is only acceptable under the special conditions of price guarantees and hill land subsidies of the British agricultural support system.

In the wet tropics agriculture is virtually restricted to perennial plantation crops such as rubber, cacao and oil palm, to shifting cultivation, and to rice culture. The tree crops have the advantage that the ground is covered by a permanent vegetation, and it is possible to dispose of excess water by controlled run-off. Soils are poor but root range is extensive, and where the returns are adequate, fertilisers can be applied readily. Shifting cultivation in the wet tropics is commonly practised with root crops or bananas as the staple. This is low production, exploitive agriculture at poor nutritional standards, both for the human population and for the crops by which they are supported. Rice is in a class apart. Since it grows in slowly moving water, large excesses of water can be managed and disposed of without undue difficulty. Nevertheless even rice has only a limited adaptability to the climatic limitations of the wet tropics. The long term consequences of an excess of water supply over water use are not to be avoided. The old soils in the wet tropics are heavily leached and hence poor in nutrients. The rice civilisations of these areas are very largely confined to young soils of better nutrient status. These are of two kinds: alluvium that has gained in nutrients from the leaching and erosion of higher lands, and volcanic soils in which fresh nutrients are steadily released by the weathering of new minerals.

The importance of an understanding of this relationship between water supply and water demand in the wet tropics is that these areas are commonly regarded as constituting a large reserve of as yet unexploited land on which the overspill of the human population could well be accommodated. Not only are they poor and agriculturally unprofitable: the prospects of amelioration so that they could support an extensive food crop agriculture are very small. Erosion and leaching under arable agriculture would be devastating. Even under perennial crops, losses of fertiliser nutrients by leaching are heavy. Those areas suited to rice production are almost all in use already. Only in the field of perennial grass cultivation is there a real prospect of further exploitation. But sugar-cane is already produced in excess of human needs, and intensive grass production for livestock production in the tropics is still in the experimental stage.

The great areas of agricultural production are those in which there is a fair balance between supply and demand for water. Under rainfed conditions these are regions receiving roughly between 1500 mm and 500 mm of rain a year. For irrigation, they are largely areas in which the rainfall is less than 500 mm a year, and supplementary water is available from rivers or from underground aquifers. In these areas of favourable climate the constitution of the soil is the dominant factor in farming success.

Soil is the storage medium for water and other plant nutrients, and under natural vegetation an equilibrium is built up under which water, usually as rain, sometimes as flood, is accepted by the soil and stored until it is used. Excess water either stands on the surface, percolates through to the water table, or runs off the surface. Standing water gives rise to swamp conditions, agriculturally intolerable except for the rice crop. Percolation leaches soluble nutrients and results in impoverishment. Run-off from stable vegetation does little damage and gives rise to clear streams. Where there are breaks in the vegetative cover, there is usually some soil loss, giving turbid streams in time of flood, and erosion. Erosion, however, is a natural phenomenon, not primarily the result of agriculture. It is the process by which landscapes have been formed from the beginning of time. Every sedimentary rock in the world is the product of erosion somewhere, at some time. Where the rate of erosion is so slow that soil regeneration keeps pace with it, no harm is done. Where it is faster than this it results in soil deterioration. And one of the most important consequences of cultivating a soil is to make it unstable and vulnerable to accelerated erosion. Hence one of the greatest problems of agriculture is the control of water falling as rain on cultivated land. It should percolate in to the limit of the capacity of the soil to absorb and store it within root range of the crop. In excess of the water holding capacity of the soil it runs through and leaches out soluble nutrients. Agricultural practice should be designed to run this excess off the surface without losing soil with it.

TABLE 1. *Soil loss in one rainy month at Namulonge, Uganda*

Type of storm	Number of storms	Rainfall (mm)	Run-off (mm)	Soil loss (metric tons per hectare)
Heaviest storm	1	82.5	35.0	17.5
Second heaviest	1	56.5	23.5	13.5
Other storms with measured losses	7	93 (aggregate)	15.0	8.5
All other storms	13	74 (aggregate)	nil	nil
Totals	22	306	73.5	39.5

The erosion problem is to a large extent a problem of rainfall variability. A large proportion of all erosion losses occur in a few violent storms. At Namulonge in Uganda, where the rain is not violent by tropical standards, in an exceptionally wet month 22 storms behaved as set out in Table 1. Thus of a total soil loss of nearly 40 tons per hectare in the month, 17.5 tons (44 per cent) went in the biggest storm, and 31 tons (78 per cent) in the two biggest storms. Effective soil conservation is very largely a matter of taking precautions against the rare event, and it is therefore important to know how serious the rare event, such as a storm of this severity, may be.

In considering the significance of erosion in agriculture, it should be borne in mind that the best natural soils are those developed on alluvium, and which have therefore been generated by erosion. Alluvium may be of any age. There are alluvial soils from such ancient erosion episodes as gave rise to the heavy clays of the Sudan Gezira, as well as the recent deposits of erosive processes that still continue. The gains to agriculture from these are illustrated in the Nile valley by the maintenance of the fertility of the soils of Egypt over 5000 years of farming use, and in Greece by the development of the fertile plain of Arta by little more than 1000 years of silting of the gulf of Amvrakia. Less valuable, but still productive, soils are generated in the deltas, inland or coastal, of ephemeral torrents from the arid mountains of Arabia and Ethiopia.

Against these gains to agriculture must be set the losses from erosion in the upper parts of river catchments, and these losses can be very great. Nevertheless, they should not be regarded as a consequence of agricultural mismanagement. Only erosion accelerated by farming practice can be blamed on agriculture. Indeed most mature farming systems have evolved techniques whereby the rate of erosion is moderated, and in many places is reduced below the rate at which it would proceed under natural vegetation.

Erosion as a natural force must be assessed in the wider context of the contribution of rainfall to the weathering process by which soils are generated. On the ancient land masses of the world, which have never been under the sea, these weathering processes have given risen to soils often of very great depth, and generally of very low nutrient status. By erosion they have been reduced to low relief peneplains, and by leaching most soluble material has been removed. Such plant nutrients as remain are locked up in the living vegetation or its dead and decomposing litter. These are the soils on which peasant communities in Africa are able to exist only by shifting cultivation. The small nutrient supply is mobilised by cutting and burning the vegetation. It is then rapidly dissipated by the leaching that goes on under cropping, and a shift is enforced by the rapid decline in productivity. Land of similar low relief and low nutrient status is characteristic of much of Australia, but in contrast to Africa, the Australian climate has been so dry that there has been only very limited weathering. Soil profiles are in general very shallow, and the deep, weathered and leached soils so common in Africa are rare. In Australia, with the capital resources of a western type community, and with an outstanding research service, a settled agriculture has been established. It is based on large farms and soil improvement by fertilisers and leguminous forages, and has given rise to production of world importance, in spite of the handicap of poor soil and very uncertain rainfall.

Richer soils are to be found on geologically younger rocks. In

Africa where granites or volcanic rocks are intrusive in areas of ancient geology, the change in soil fertility over a short distance can be dramatic. Weathering processes have not gone so far on the younger rocks, and new nutrients become available by decay. A great range is to be seen in degree of weathering. An advanced stage, but one still yielding useful agricultural soils, is illustrated by the soil catena associated with granite outcrops in Sukumaland in Tanzania. Here, coarse sandy soils of fair fertility are found below the granite kopjes, with an impervious, badly draining soil in the middle levels, and a heavy alkaline black cracking clay, or mbuga, in the valley bottoms. The extremely young stage in soil development may be illustrated by the volcanic ash soils of St Vincent in the West Indies, where soil differentiation is only beginning, and if there is very heavy soil loss by erosion, a new fertile layer can be generated in a few years by the establishment of leguminous plants. The chief deficiency is nitrogen. Other nutrients rapidly become available from the decaying volcanic ash.

The soils of northern Europe are distinct in that they are recent, dating from the last Ice Age. They are diverse, both in physical structure and in nutrient status, but they have not been exposed to the enormous periods of weathering of so much of the soils of the tropics. The very light soils of the sand lands are limited in their agricultural potential by the smallness of the water reserve they maintain. A week's dry weather in the summer brings a crop into a period of water stress, and yield reduction follows in a considerable proportion of years. Usually, nutrient poverty goes with poor water holding capacity because the rainfall freely draining through removes the soluble nutrients. So a light soil is generally a thirsty one and a hungry one also. Britain's light soils were the first exploited, and in some measure they were degraded by early farming. Not many went out of cultivation, but there are areas, for example on the East Anglian Breckland, that were only cultivated intermittently, and provide a temperate region counterpart to the shifting cultivation of the tropics. For the most part, however, soils once brought into agricultural use were so maintained permanently. The deeper, heavier soils, more retentive of moisture and nutrients, carried a heavier forest cover and were brought into use later. Indeed right up to the middle of the 19th century, fresh land was brought into use in England as technology improved. First the improvement was in means of clearing and cultivating, then in drainage, fertility conservation and buildings, and finally in the amelioration of soil deficiencies by fertilisers.

All this involved a process of change in the soils of Britain that can best be described as domestication. It is not usual to think of soils as 'wild' or 'domesticated' but in a very real sense the establishment of a highly productive agriculture involves the domestication of the soil as much as of the crop plants and livestock that live on it.

The agricultural soils of Western Europe have been under cultivation for so long that they have long been fully domesticated and there are few 'wild' soils in any way climatically and topographically similar with which to compare them. Occasionally, as when an old oak wood is cleared and brought into cultivation, one can see at least some stages of domestication in progress. Drainage is important and of little significance in the wild. The pH is a major factor in the establishment of good cropping conditions. Structure, nutrient status, organic matter content, and aeration are all factors in the specification of a soil that must be brought to certain standards before good crop production – or grass production – can be ensured. Some of these conditions are met in most wild soils. It is balancing them all for crops that leads to domestication. Moreover, even if all the changes necessary can be specified, and all the ameliorating treatments can be applied, it will still take some years for the character of the soil to change, and the agricultural adaptation to appear.

The changes involved in domesticating a soil are more readily seen in the establishment of agriculture in new countries. A good example is the planting of cacao on forest soils in Trinidad. Control of ground vegetation and drainage in particular initiated long continued and profound changes in the soil. These were not always to the good, but it simply is not possible to maintain a forest soil, with the high and variable moisture levels characteristic of it, and with the large accumulations of slowly decaying organic matter, under a tree crop where all the trees had to be given favourable conditions for growth. Similarly, the organic matter status and the distribution of nutrients in the soil profile change rapidly with the establishment of arable cropping in place of elephant grass in Uganda. The whole water regime of a region may change, as was demonstrated by Bunting (1961) in the orchard savannah of the Sudan rainlands. The open orchard savannah was adapted to making full use of the short rainy season by trees which had established root systems ready and able to take up water, right from the beginning of the rains. Annual crops cannot be sown until the rains have begun, and they use little water until they have developed a large leaf area. Thus, to quote Bunting: 'Under open woodland with annual grass and herbs on the heavy clays at Tozi the seasonal penetration of moisture does not usually extend below 12 inches, whereas under annual cropping it usually reaches four feet, representing up to 10 inches of available water, all within root range.' In all these cases it is no use trying to maintain the 'wild' conditions, even if they were good. Nothing but the development of a new balance will make possible continued agricultural use of such lands.

Land with an agricultural potential is land in which a new balance can be struck between climate, soil, and vegetation that is useful to man. The latter may be grass or legumes for forage, arable crops, orchards and other tree crops, or forest. These will create their own balances, which will differ according to the species used and the climatic and soil circumstances, and they will only to a limited degree be interchangeable. Land on which no such balance can be established cannot be regarded as agricultural land. It may be exploitable, but unless a permanent and stable and self renewing system of use can be devised, it cannot be classed as agricultural land. Much agriculture has been practised on an exploit and abandon basis. The banana industry went through the mountains of Jamaica like a plague. Coffee cultivation has done the same in both Jamaica and Colombia. The fertility accumulated under the stable forest cover was exploited and dispersed, and when the land was exhausted the plantation crops faded out. Often grass invaded, and grass is food for cattle, and cattle replaced the plantation crop. But it does not end there. The grass goes and the soil goes and sooner or later the whole slow process of rebuilding under natural cover must begin again.

Some such exploitive systems run down to a fairly stable base level. The exploitation of the limestone ranges of Epirus in

Greece has done so. Originally covered with oaks, forest in the better areas, and doubtless scrub in the areas with least soil, they have been denuded of their trees by cutting and browsing. Much soil has been lost, but an oak scrub cover persists which provides browse for a large goat population. In the very long term, further deterioration will occur because oak does not fruit as scrub, so replacement of any scrub that dies is only possible in the vicinity of surviving trees. But oak being very long-lived, further deterioration is so slow as to be hard to detect.

The three examples from Jamaica, Colombia, and Greece, are all examples of the problems of agriculture on very steep land. In all three cases, there is good reason to believe that stable and self regenerating systems of land use could be devised. With such perennial crops as bananas and coffee, soil conservation measures can greatly prolong the life of the crop. It may be that bananas, involving heavy transport costs for bulky and perishable produce, should not be cultivated on steep slopes, but coffee certainly can, and in Colombia as a national policy, 25 years in coffee and 75 years in timber might well make a stable, permanent form of land use. On the Mediterranean scrub-covered mountains, it is clear that the only form of land use is by browsing animals, and the goat is ideal for the purpose. A limited amount of gully stopping, the protection of oaks so as to provide an adequate scatter of seed bearers, and browsing control and resting could well increase the available browse very substantially.

There are areas where such stabilised systems are not possible. In St Vincent, very steep volcanic ash slopes are cultivated in food crops, with a little cotton and arrowroot. Erosion losses are enormous. Conservationists have argued that the land should be retired from agriculture and conserved. But for what should it be conserved? It can never be used for anything but the type of peasant gardening that is now practised, and no amount of conservation will create conditions under which it will be any more stable than it is now. So it is at least arguable that if people want to use it, they may as well. They are getting something now, and they are doing no more damage to the resources of the future than a man who exploits a coal seam or an oil well.

The domestication of soil for high productivity agriculture depends on choosing areas within acceptable limits of topography and climate, and amending the soil on them to suit farming needs. This can be done, but only within limits. Peaty swamps can be drained to make fertile black fen farms. Acid sands can be limed and heavy clays can be drained. But they are still peats, sands or clays, and their response under agriculture parallels their behaviour under natural vegetation. Nevertheless, to bring about this degree of improvement, a great deal of capital has been put into the soils of all advanced communities. The original clearing from forest involved cutting and burning or removing timber, and stumping the land. Doubtless crops were grown between the stumps in the first instance, but stumps have to be removed before even oxdrawn implements can be used. Stump fences can still be seen in Canada, tree roots set on edge in a row making a very satisfactory barrier. All the trees were cut some years before their roots were extracted. Stones are a problem in many lands. One can see walls six feet wide in an area of Sweden where the glaciers left quantities of stone, and the stone walls in Britain also serve the purpose of stacking stones out of the way as well as of fencing the land. Drainage, access, shelter for man and beast have all to be provided and moreover the great irregularities of 'wild' land have to be smoothed out through expensive cultivations. All these are included in the capitalisation of new land. In Europe, they were put in centuries ago and have been written off. In the developing countries this capitalisation has still to be undertaken. It is slow, but it is rewarding. A good agricultural soil is an asset to mankind and a testimony to the skill of the man who made it.

CHAPTER 3

CROPS AND STOCK

Considering the enormous number of plant and animal species available, those that have been domesticated and brought into a symbiotic relationship with man are very few. Man has experimented with a somewhat wider range. Godwin (1965) concluded on botanical evidence from Danish bog burials of the Iron Age that *Chenopodium album* (fat hen) and probably also *Polygonum lapathifolium* were cultivated, and that some other weeds of fallows were at least systematically collected and separately stored. Where such experimental domestications failed and the plant was abandoned, it still persisted as a weed of cultivation, and plants like *C. album* and *P. lapathifolium* remain as weeds in spite of 5000 years of active discouragement by man.

As the successful domesticates were progressively improved, the prospect of bringing in other wild plants in competition with them receded. Modern agriculture is, in fact, firmly based on plants and animals which were domesticated before the dawn of history. Only in response to new needs has the domestication of new species from the wild been successfully carried out. The most important recent domesticate is the *Hevea* rubber tree, brought into cultivation only a century ago, and already showing the characteristics of varietal diversity and response to selection for productivity of a true crop plant. More recently, the intensification of livestock production in new agricultural areas has led to extensive plant exploration, and the introduction and establishment of new cultivars of forage plants. In Australia in particular, both grasses and leguminous species have been introduced into cultivation, and successfully improved by selection. Thus there is no bar to the extension of the list of domesticated plants, if new needs are felt that are beyond the limits of performance of those now in cultivation.

Similar considerations apply to the domestication of animals. Hitherto, agricultural needs have been adequately met by the species already tamed. Recently, some new needs have arisen. A wider range of laboratory animals is required and such cage species as the laboratory rats and mice and hamsters are recent

28

acquisitions from the wild. One can foresee further advances in this direction, as for instance in the establishment of a domesticated race of the *Rhesus* monkey. In agriculture, however, there have so far been no new animal domestications. With the intensification of African agriculture and the increased interest in dietary balance, the domestication of antelope species has been advocated for the production of meat. This would seem a sound and perfectly feasible enterprise, well worth undertaking.

Since all the major crop plants and domesticated livestock were established in agriculture in prehistoric times, the course of improvement in the early stages must be inferred from archaeological data, and from the evidence gained from genetic and cytogenetic studies of cultivated species and their wild relatives. For wheat and barley, the primary cereals of the Middle East, Zohary (1969) has recorded the existence in Palestine today of extensive stands of wild *Hordeum* and *Triticum*. In his view, yields from these stands would be much the same as yields from primitive cultivars under local primitive agricultural conditions. It may be supposed, therefore, that the activities of the first farmers were devoted to extending the area of the natural stands, and it was in these circumstances that the natural forms underwent genetic change.

The establishment of storage, and the extension of area by cultivation and sowing, resulted in a radical change in the selective forces acting on the species. Characters that favoured seed dispersal and natural survival ceased to be advantageous and often became positively deleterious. Here appears the first indication of the characters on which agricultural archaeologists such as Helbaek depend as criteria of the difference between wild and cultivated plants that are used for food. In wild forms of the cereals the ear is fragile and breaks up into spikelets, and the grains remain firmly enclosed in the glumes. The spikelet is the means of dispersal and of placing the seed in a suitable position for germination. In a crop plant this dispersal mechanism results in substantial losses by shattering at harvest, and in the need for special treatment such as parching, to free the grain from the glumes. So if storage and sowing was assured, loss of these characters by the emergence of tough rachis and free threshing mutants was advantageous. These are found established throughout the cultivated cereals very early in agricultural history.

Other crop plants have undergone parallel changes. Bean and *Brassica* pods in wild plants split and scatter their seeds on ripening. Cultivars have been bred which do not shatter. Moreover, the same process is going on with crop plants still in a relatively unimproved condition. In *Sesamum*, for example, with pores that open in the top of the capsule and shed the ripe seed, much work has been done – not yet fully successfully – in breeding a non-opening gene into commercial varieties. Similarly, closed boll and 'cling' characteristics offer prospects of improving the suitability of cotton varieties for machine harvesting.

Next in order in the evolution of cultivars is the change in plant habit. Cropping, and in particular high production cropping, involves regular tillage, and good soil management is easier under annual crops than under perennials. Moreover, where the seed is the useful product, a higher proportion of useful material to leaf and stalk is obtained if the build up and maintenance of a large vegetative structure is obviated. So in terms of useful dry matter production, wheat is better than apple trees, annual cotton is better than perennial, and short term (five months) sorghum than long term (ten months) sorghum. For the same reason of improved productivity, semi-perennial scrambling beans have been replaced, first by pole beans (single stem) and more recently by dwarf bush beans. Maximisation of useful material has gone on also in crops of which vegetative parts are of value. Flowering and seed production have been reduced in sugar-cane, and true seed production in potatoes. In vegetables the useful product has been increased by the manifold hypertrophies of the *Brassicas* – cabbage, cauliflower, sprouts, turnips, etc. – and lettuce and beet to name but a few.

Major changes in chromosome complement, leading to the establishment of new and distinct cultivated species, have gone on in some crop plants. The allopolyploid series in wheat and oats are well known and have been extensively studied. In them allopolyploidy has been associated with changes that improved their suitability to man. Nevertheless, major changes in the chromosome complement are only occasional incidents in crop plant development. The improvement under cultivation of barley, maize, and rice has been no less striking than that in wheat and oats, and has gone on without polyploidy. In sorghum polyploidy occurs but is not associated with the advance of the crop as a cereal. Polyploidy may be incidental to the evolutionary progress of a crop plant. It is not necessarily, or even commonly, associated with it.

Evolutionary changes under domestication have been less radical in animal than in plant species. Rarely has anything so new to nature as the maize ear, or even the tough cereal rachis, been evolved in a domestic animal. This is one reason why the beginnings of animal domestication are so hard to detect in the archaeological record. Perhaps the most radical evolutionary change was the development of sheep's wool to the point where it became useful as a textile raw material (Ryder, 1969).

Characters of value in domestic livestock were docility, growth rate, ancillary production (wool, milk, eggs), working characteristics, and conformation for meat, all of which with the exception of wool involve only small morphological changes in the animal. Hence improved stocks from time to time re-establish themselves in the wild – cattle on the Falkland Isles, goats on semi-arid islands, even poultry on Little Tobago in the West Indies. As the characteristics associated with high productivity are developed, however, the interdependence of man and livestock is intensified. Thus a modern dairy cow produces so much milk that survival would be impossible without the dairyman to relieve her, and modern high egg producing poultry strains would die out without incubators to replace the lost brooding habit.

High productivity in both crop plants and livestock is a matter of environment as much as of genotype. The history of livestock improvement in Britain illustrates this. The improvement in beef cattle and in sheep associated with Bakewell and the founders of the breed societies was only worthwhile when improvements in feeding had been established. Thus the establishment of the turnip and clover crops, and the building of steadings in which cattle were housed in winter were as important as selection and breeding in the great advance in livestock performance. In the same way in recent years, the great increase in milk production in the British dairy industry derives equally from improved standards of feeding and from improved breeding through progeny testing and the use of proven sires through artificial insemination.

In some ways the success of the late 18th and early 19th century stockbreeders impeded further advance in the 20th century. Breeds were closed and crossing was anathema. So while plant breeders were exploring the variability available to them by hybridisation and selection, animal breeders were practising selection in small, closed herds. This led to some isolation of the pedigree breeders from the main stream of livestock production. In both beef and sheep, meat production has come increasingly from crossbred stock, to which the pedigree breeds contributed only as parents of the hybrids. The dairy industry has had the advantage that the Friesian breed is a recent introduction and very many of the British Friesian dairy herds are the result of recent grading up to Friesian bulls and hence contain in their population some genetic diversity from other breeds. Moreover, introductions from Holland, Canada, and South Africa have also broadened the genetic base of the breed. Progress under selection in these circumstances has been good, and doubtless will continue. In other sectors of the livestock industry, the gene pool was fragmented among small, uniform breeds, and the objectives of selection were primarily for the maintenance of an established standard. This was an entirely static policy quite unsuited to a period of rapid agricultural change. Hence, the campaign run by agricultural scientists and leading producers, to introduce into Britain the French Charollais beef cattle, the Swedish landrace pig, and the Finnish landrace sheep. These have broadened the genetic base of British livestock breeding, but more important, they have greatly increased the extent to which stock of hybrid origin is used in breeding and selection on British farms.

In developing countries livestock improvement has come first from veterinary medicine, particularly preventive medicine. Very great gains have been made throughout the developing world, but especially in Africa. With the large wild ruminants of Africa, there is a great range of pests and diseases that move onto domestic cattle as soon as they make contact with them. Control through preventive medicine has been with stock, as with humans, a major cause of population increase, and this has gone on to the point where the stock population presses hard on the food supply. Custom does not change as fast as veterinary medicine advances, and stock keeping practices suitable under the old conditions do not allow of a large enough offtake of meat under the new. In Africa this is an adjustment that is steadily coming about wherever the cattle keepers are in sufficiently close touch with market demand.

The need for animal protein is such that it is urgent to press on with the next stages of animal improvement, which are improved nutrition, and improved breeding. That the latter is of little use without the former is indicated by experience both in Australia and in Uganda. Steward, speaking of beef ranching in northern Australia, said that from the end of one rainy season to the beginning of the next (say seven months) an 800 lb steer will lose 200 lb in weight, though he has as much dry grass as he can eat. There is an enormous amount of meat and some milk to be gained in developing countries simply by giving enough digestible supplement to cut out that dry weather loss. Obviously, genetic improvement will not be worth while until losses of this kind have been eliminated. So the first step in improvement must be the provision of adequate feed supplies the year round.

TABLE 2. *Response of Teso Zebu milking cows to supplementary feeding*

Supplement	Milk yield (lb)	Days in milk	Weight changes during lactation (lb)
8 lb sorghum 2 lb cotton seed	1193	278	+16
Cooked cassava	1236	274	+15
Nil	1177	274	−11

Joblin (1966), working with a milking herd of Teso cattle in Uganda, showed that when adequate feed was given, local cattle immediately reached a production ceiling. He provided a basic 'no supplement diet' of adequate grazing, straw, stover, or other roughage the year round, and then compared supplementary rations as set out in Table 2. Thus, just as genetic improvement is unprofitable until sound nutrition levels can be assured, so a high plane of nutrition is unjustified until the genetic potential has been improved.

Milk production is a line of improvement of great importance since, in the supply of milk to urban communities, farming people have an opportunity to sell a high value commodity to a sector of the population able to pay for it. Milk poses problems of hygiene, storage and transport that limit its place in a farming system to those areas with rapid transport to an urban market. Cattle production for meat is not so restricted, and once the limit of seasonal inadequacy in the food supply has been overcome, breeding for rapid growth and good fleshing characteristics becomes worthwhile. As with milking capacity, livestock types in tropical countries vary greatly in their suitability for meat production, and substantial advances can be made by selection.

Cattle breeding the world over has been revolutionised by the technique of artificial insemination. In western countries the proportion of males necessary to maintain a population has been greatly reduced, and advanced selection techniques are used to ensure that the best are chosen as sires. In developing countries the very small numbers of elite stock, and the ready availability of males from temperate regions through semen importation, have made crossbreeding with high production temperate breeds the most effective means of breed improvement. Improvement in this way is rapid, but the use of imported male stock is no substitute for a breeding policy. Tropical breeds of cattle have substantial advantages in disease resistance and climatic tolerance, and probably also in the efficiency with which they convert low grade forage. The aim of breed improvement should be the establishment of productive local breeds, and not merely the transfer of temperate breeds to tropical countries.

In this discussion attention has been paid primarily to cattle, partly because of their very great importance in world agriculture, and partly because of the extent to which the principles and achievements of breed improvement are illustrated by them. However, there are principles that have been exploited in the improvement of poultry that must also be considered. Poultry have a short life-cycle and a high potential multiplication rate, and they can be kept under closely controlled environmental conditions. In these circumstances it is possible to exploit heterosis by the use of hybrid stock for production purposes. Hence virtually the whole of the vast poultry industry, both for eggs and for poultry meat, of North America and Western Europe is based on mass produced hybrids derived from very highly selected inbred stocks held by major breeders. Local adaptability is no problem, as environmental control is very close. In the developing countries it is possible to bring in the whole of this technology by setting up controlled environment buildings, following the feed specifications, and importing chicks by air freight. Such a policy is acceptable as an interim measure to meet the needs of growing urban areas, but in a policy of improving local agriculture for the benefit of all farmers, the improvement of local poultry stocks is important and local stocks can best be improved not by hybrids, but by well bred, productive, self propagating strains.

As with livestock, so with crops, increasing productivity has come from a combination of environmental and genetic improvement. In Britain in the 1920s and 1930s, when agronomic change was very slow under the agricultural depression, plant breeders were content if small improvements could be achieved. In the 1950s, when the demand for home produced food stimulated a great improvement in agronomic practice, the breeding of new varieties capable of exploiting the new conditions led to yield increases of the order of 50 per cent in a few years. The genetic changes by which these yield advances were achieved are probably only in small measure changes in intrinsic productive capacity. Trials in which long established and modern varieties have been compared rarely show large yield differences in favour of modern varieties unless the environmental conditions penalise the old varieties. Thus, if the old varieties do not lodge under high fertiliser treatment and are not seriously damaged by diseases to which the new varieties are resistant, they may compare not unfavourably. In farmers' fields, however, these are the factors of the new agronomy that create a serious hazard for the older types. Having met and dealt with these problems, breeders in advanced countries are now deeply concerned with studies of the physiological nature of the synthesis of yield, in the expectation that the next generation of varieties will be bred for intrinsic yielding capacity.

The achievements of crop botany in tropical countries have been greatest in the improvement of tropical crops grown to meet the demands of temperate regions. The exchange of such agricultural products for western industrial goods has been a major factor in development, and these crops have been very extensively grown outside their former geographical limits. The modern cotton industry, for instance, dates from the establishment in the United States of the Upland crop, which was derived from cottons of Central American origin. The Upland cottons are now grown very widely in Africa and Asia as well. Cacao from Latin America is produced in great quantity in West Africa. The world production of rubber, which is of Latin American origin, is centred in Malaysia and Indonesia. Coffee of African origin is extensively grown in Latin America, and sugar-cane and bananas from South East Asia are now produced throughout the tropics. In this situation the agricultural botanist has had the opportunity to improve the response of the crop to the new environment both by breeding locally adapted varieties and by devising pest and disease control measures to meet the new circumstances. In all these crops original introduced races are giving way to, or have been superseded by, locally bred varieties selected for suitability to local conditions of climate, soil and husbandry. Much of this work has been based on plant exploration and introduction, since the original introductions were often limited in quantity and in genetic diversity. Some has been very long term, as with the seedling selection of cacao and the clonal selection of rubber. Some crops, such as the banana, have presented major genetic and physiological problems in devising any effective breeding technique.

Some of the greatest successes have naturally been with shorter term crops such as sugar-cane and cotton. Breeding in sugar-cane began about the turn of the century, with the discovery in Barbados that sugar-cane produced viable seed. The culture and selection of seedlings was an immediate and spectacular success. Seedling canes greatly superior to the parent material were grown in large numbers, and a sequence of new and superior varieties put new life into the industry. The new varieties were, however, unexpectedly ephemeral, and were said to 'run out' or 'degenerate' in a few years. This was thought to be something to do with continuous vegetative propagation. In fact this belief was remarkably close to the truth, though the effect of vegetative propagation was not to cause a genetic degeneration, as was first supposed. Degeneration was due to infection with virus diseases, which were carried on through the vegetative planting material. Thus the freedom of a seedling from virus infection accounted for a large part, but not all, of the improvement gained.

The breeding of seedling canes was soon undertaken in Coimbatore in South India, and rapidly became a necessary part of the industry in all important sugar-cane growing countries. For many years the industry has depended entirely on the seedling canes produced by breeders, and the production of sugar has been greatly increased thereby. The simple early breeding systems involving selection among great numbers of seedlings have given place to more sophisticated programmes, such as Dutch 'nobilisation' work, involving controlled polyploidy, and the Indian work of crossing with other species of *Saccharum*, and even *Sorghum*, to give canes tolerant of the hard conditions of monsoon India.

In cotton, breeding improvement began with the selection of annual strains neutral to day length for cultivation in the southern United States, where short day types did not fruit before the onset of winter. It went on all over the world, mostly but not entirely with New World cottons, which were very widely distributed and have replaced the Asiatic species in many areas. Selection of locally adapted types has involved adjustment to the length of the season and the day length regime, and disease and pest resistance. Commercial success has depended substantially on quality considerations, especially where high transport charges put a premium on high value per unit weight.

In cotton breeding in Africa, pest and disease resistance have been given particular attention. Parnell and his colleagues (1949) first showed that leaf hairiness was directly responsible for resistance to the jassid pest. Their hairy selections made in South Africa were so greatly superior to non-hairy stocks that they were very widely grown. Breeding thereafter led to the establishment of a range of locally adapted hairy types in the various African cotton growing territories. Knight's (1946) study of breeding for resistance to bacterial blight in the Sudan was equally significant in the improvement of New World cottons for growth in Africa. He was particularly concerned with Egyptian (*Gossypium barbadense*) cottons, which are virtually confined to the Nile valley. Nevertheless, his genetic stocks have been used in breeding elsewhere, and the understanding he gained of the genetic control of resistance, and of the sources of resistant material, have been of very great value in the elimination of the disease by resistance breeding in Upland (*Gossypium hirsutum*) as well as Egyptian (*G. barbadense*) cottons.

Cacao breeding is a classic case of improvement dependent on plant exploration and introduction. Cacao research workers in the West Indies deduced that the centre of origin of the cacao crop was in the forest areas of South America drained by the great rivers of Colombia, Venezuela and northern Brazil. Expeditions were sent to collect material, in the first instance in the hope of finding stocks resistant to the 'witches broom' disease which was causing serious damage to the Trinidad crop. Posnette, working in Ghana, realised the potential of the new collections for widening the genetic base of the West African crop. Cacao is a tree crop with a long life-cycle, but even so, it has required less than 50 years work to establish new highly productive cacao strains in both Trinidad and Ghana, and resistance to 'witches broom' is now available in Trinidad, and tolerance to the virus complex causing swollen shoot in Ghana.

Banana breeding illustrates in an extreme form the problems facing the breeder of plants that are normally seed sterile. The edible banana is a parthenocarpic fruit, in which seed formation never takes place. Cultivars are diploid or triploid, and parthenocarpy is reinforced by irregularities in the chromosome complement such that the reduction division never results in the formation of a balanced gamete. Wild inedible but fertile diploid bananas produce an abundance of normal pollen and these were used to pollinate large numbers of female flowers on edible bananas. Very occasionally a good seed was formed, and these seeds were found always to be the result of the fertilisation of an unreduced egg. The chromosome number of the resultant seedlings was consequently higher by one complete genome than the female parent. They were however sterile and parthenocarpic. Up to the tetraploid level, an increase in chromosome number is compatible with good vigour, so there arose the possibility of breeding new types carrying one genome from the wild species, but retaining the sterility and parthenocarpy of the edible parent. With this information Dodds (1943) propounded a breeding system on which parthenocarpic edible bananas could be improved. He proposed that the characters desired in the edible type should be bred in the seeded inedible diploids, and when a superior stock had been produced, its whole genome should be added to that of the edible clone it was desired to improve, by extensive pollination in order to produce seed from unreduced eggs. In this way new and improved clones of edible bananas have been bred.

This type of work has led to the establishment of sound stocks of the main cash crops in most of the areas in which they are grown commercially. Improvement in tropical food crops has not been so great. For most of the colonial period in Africa, food supply was no great problem. The world's food markets were over supplied by the massive production of the new lands in the Americas. In the rest of the developing world, once famine reserves had been established, the limited research resources available were devoted to cash crops. That change was possible if change were needed was evident. When, for instance, Doggett showed that there were advantages to be gained by growing short term good storing sorghums in East Africa, he had no difficulty in breeding a suitable type.

India, with a large population to feed, has a long history of plant breeding work on food crops. Improved varieties of all the major food crops have been available over the past half century. Nevertheless, the influence of breeding work on the crop populations of the country was only exerted slowly, in the absence of an organised seed industry, and under adverse economic circumstances. Food supply was largely a matter of family production for family use, and incentives for increased production were small while the disincentive of the great agricultural depression dominated the peasant's attitude. It is with the great increase in population, and especially urban population, of the last two decades, that the improvement of Indian food crops has become a pressing problem. Increased crop production came primarily from increases in the area of irrigated land and from increases in fertility level from the use of fertilisers. Following these agronomic improvements fertiliser responsive races have been bred of all the important Indian cereals to exploit the new potentials, and the increasing demands of the urban market. These new races are high potential varieties, and not high yielding varieties. Their value is in their responsiveness to high fertility. Indeed, at traditional Indian fertility levels they are inferior to traditional Indian varieties. For wheat, Kohli's (1969) data may be quoted. He compared the yields of an Indian bred variety, Punjab C.306, and two American dwarf varieties, Lerma Rojo and Kalyansona. Trials were carried out at 11 and 12 centres in three successive years, at two fertility levels. The low fertility plots received no fertiliser nitrogen and the high fertility plots were dressed with 135 kg nitrogen per hectare. Mean yields in quintals per hectare are given in Table 3.

TABLE 3. *Yields of three varieties of wheat (in quintals per hectare) at two levels of N fertiliser over three seasons and 11 or 12 centres*

Year	1964–5		1965–6			1966–7		
No. of centres	11		12			11		
Variety	C.306	Lerma Rojo	C.306	Lerma Rojo	Kalyan-sona	C.306	Lerma Rojo	Kalyan-sona
No N	23.2	20.1	20.1	18.2	18.9	21.2	18.6	19.2
135 kg/hectare N	33.6	39.7	31.0	33.3	33.6	25.7	26.1	31.3
Percentage increase due to N	42	98	54	83	78	21	40	63

Both Mexican varieties yielded less than the Punjab variety – by about 10 per cent – at the lower fertility level. The increase in yield due to fertilisers was about twice as great with the Mexicans as with the Punjab variety, and with fertiliser they yielded 5–20 per cent more than the Punjab variety. The varietal difference is much less than the advantage experienced in general cultivation. The greater advantage reaped by the cultivator may be ascribed to those other factors of standing power and reliability at high fertility levels that tend to be obscured by the care and attention necessarily given to precise variety and fertiliser trials.

The recent breakthrough in wheat and rice production in

India has been achieved with varieties brought in from central breeding stations in other countries. These two spectacular successes have led to a belief that agricultural progress can be stimulated by large central crop breeding stations, organised on a world basis. However, an assessment of the limitations of the new varieties emphasises the importance of local rather than centralised breeding work. Both the Mexican wheats and the Philippine rices have quality characteristics that are not acceptable in Indian cooking. Quality in food is essentially a matter of local preference, and can only be effectively dealt with in breeding work at the country level. The rice varieties have also proved susceptible to pathogens to which most Indian rices are resistant. The Mexican wheats have not hitherto suffered to the same extent, but there is the very real risk that they may break down to new races of rust. Moreover, there are hidden hazards in the introduction of varieties bred in other parts of the world. Pests and diseases to which local varieties have become so resistant as to make them of no importance, may develop to epidemic proportions on introduced material. The first introductions of Upland cottons into India a century ago failed because the jassid pest emerged from obscurity in this way, and in recent times, hybrid wheat varieties with a Kenya parent failed through attack by *Alternaria triticina* (M. V. Rao, personal communication), which had not been known before as a disease. There are less tangible reasons also for regarding local breeding using local stocks as in the long run essential for continued success. There is no doubt of the importance of local adaptation to local climatic circumstances, and since these cannot be simulated, the only way to ensure that they are taken into account is to breed under the local conditions.

The nature of the achievement of the central breeding stations is best illustrated from the history of a similar achievement in cotton breeding half a century earlier. Parnell's jassid resistant U4 referred to above was as great an advance over all other cottons in cultivation from South Africa to the rainlands of the Sudan, as the Mexican wheats are over the indigenous Indian varieties. Further improvement in the cottons of eastern Africa came, not from the distribution of further improved stocks from Barberton, but from the use of the U4 type resistance, either through hybrids or by local selection, in the improvement by local breeding, of the stocks suited to the conditions in each of the territories concerned. A second substantial advance was achieved recently with the development of the Albar stock in Uganda, and again though Albar has been widely distributed, it is used in local breeding programmes rather than being issued as a new introduction. One may conclude that a well founded breeding programme will from time to time yield a stock of such outstanding value that it will sweep through the commercial crop over a wide area. Such a stock can then be improved, not by concentration of breeding where it first arose, but by breeding within it and its hybrids to meet local circumstances. Thus, the proper organisation of breeding work is by the maintenance of a range of breeding stations to cover the major areas of production, and not by centralisation in large national or international stations. Co-ordination is best achieved by some central body such as the Cotton Research Corporation in respect of cotton breeding in Africa, or by a system of co-ordinating officers and representative working parties, such as has been set up by the Indian Council of Agricultural Research.

Plant breeding is only one of the ways in which crop production has been improved. It is of particular importance in that it is the technology most closely related to the evolutionary process. Nevertheless, this evolution has been the evolution of a symbiosis between man and other organisms, and the efficiency of the symbiotic relationship has been greatly improved by control of organisms outside it. The status of plant pathogens and pests is greatly changed by cropping practice. A disease or an insect pest more easily develops to epidemic proportions in the very large populations of a modern commercial crop than in the scattered plants or small plant communities of a natural ecology. So plant sanitation and hygiene, and control measures against major pathogens, have contributed to productivity in such fields as virus diseases of sugar-cane, cacao and cotton, and insect pests of cacao and cotton.

These are the beginnings of change in crop plants and livestock. The potential for further improvement is enormous in all agricultural systems, but particularly in those of the developing countries.

CHAPTER 4

HUSBANDRY

Husbandry is the integration of the components of agriculture into a farming system. It may be a simple form of exploitation of cleared land with a single crop, or it may involve the combination of crops and stock into a system capable of maintaining a mature agricultural soil permanently in high production. In general, there is an increase in the complexity of the system as agriculture increases in sophistication, but in modern times there is also to be seen a process of simplification and specialisation in commercial farming designed to serve modern urban communities. It is the purpose of this chapter to survey the history of the development of modern systems of husbandry.

The earliest system was shifting cultivation, and it is a form of husbandry that has had a much greater place in agriculture than is commonly recognised. Iverson (1941) likened the system that Neolithic farmers practised at the beginning of agriculture in Denmark to a cut and burn system still practised in Finland. Early English settlers in the American colonies were driven to practise a long term shift for the same reasons of low soil fertility and heavy weed growth as Africans in Zambia and the Congo practise it now. A very long term shift has been practised on the Breckland in eastern England, land being brought into use when prices were high and abandoned when it had been worked out or prices fell. So shifting cultivation must be regarded as one of the major, world wide, systems of agriculture.

No one cuts down trees to grow his crops unless he has to. Having made a clearing any sensible – and reasonably idle – farmer will use it as long as he can. So it is important to be clear on why so many farmers in so many countries do in fact shift every few years. Weed growth is one important reason. The form of shifting cultivation practised on the natural grasslands of the Sudan, known as 'harig', for instance, is designed to give good weed control, the burn being postponed until weed seeds have been germinated by the first rains.

The second important reason for shifting cultivation is nutrient status. Fertile soils are not to be found everywhere.

On infertile soils, shifting cultivation is a system that makes possible the exploitation of such fertility as there is. So long as land is plentiful, shifting cultivation provides subsistence. Given a long enough break between cultivation periods, natural vegetation remobilises most of the nutrients which in the cultivation period are leached beyond root range of the crops. The area can then be used again. Such improvement as is possible in shifting cultivation has been in the standardisation of the rest period and the encouragement of the best and most deep rooted components of the natural vegetation in the regenerative phase.

The line between shifting cultivation and more permanent – and intensive – systems of agriculture is not a sharp one. The proportion of time in crop to time in rest may be two years cultivation to 20 years rest or various narrower ratios down to the three years cultivation to three years planted elephant grass advocated, though rarely practised, in Uganda. The ratio is narrower the better the land. So there is land that has been once cropped and is thereafter useless on the sedimentary rocks of the Venezuelan mountains through a complete gradation to the continuous use of heavy land in Britain under alternate husbandry. The determining factors are weed control and soil fertility.

At the Cotton Research Station, Namulonge, in Uganda the recommended Uganda system of three years cultivation and three years rest under grass was adopted. It was apparent that by the introduction of cattle onto the grass rest, the farming system could be intensified, and if successful, the beginnings of alternate husbandry would be established. The change involved was substantial. In the first place the 'ley' of alternate husbandry is not the same as the 'rest' of shifting cultivation. A rest may involve the regeneration of coppiced shrubs and trees. Even if it is predominantly grass as in Uganda, the tall tussocky grasses such as elephant grass are unsuitable for grazing. Secondly, the establishment of alternate husbandry implies a degree of intensification that calls for convenient and rapid cultivation. So the land must be cleared of stumps to permit ploughing, by ox or by tractor, and a grass stand established that will be suitable for grazing. Radical changes are necessary throughout the system. Under shifting cultivation the vegetation of the rest period is suppressed, but not destroyed, in the cultivation phase. When the cropping sequence is completed it regenerates from coppiced stumps, and from the vast seed reservoir of the grasses and herbs of the resting phase. Under alternate husbandry not only must the stumps be removed, but the seed reservoir, which creates a constant heavy weeding problem in the arable phase, must be reduced. In advanced agricultural systems the grasses and legumes of the ley phase are sown just as the crops of the arable phase are sown. Delayed germination, an important asset to the plants that regenerate spontaneously in a shifting cultivation system, is a deleterious character to be eliminated by selection. In developing countries such as Uganda where alternate husbandry offers good prospects for intensive land use, these improvements have still to be undertaken.

TABLE 4. *Indices of wheat yield, 13th–19th centuries (Crude means per century from van Bath's Tables 2 and 3, converted to kg/hectare)*

Century	Average seed/yield ratio	Estimated yield kg/hectare
13	3.8	805
14	3.8	805
15	4.6	975
16	6.2	1312
17	5.5	1165
18	8.5	1800
19	10.75	2280

British agriculture had settled down to using continuously the better land by the Middle Ages, and British agricultural history provides a record of the improvement of husbandry on land in permanent agricultural use. In the open field system of the mediaeval manor, fertility was low. Van Bath (1963) gives estimates of yields in various European countries from the 13th to the 18th centuries. He expresses them as 'seed/yield ratios', the amount harvested per unit of seed sown, because of uncertainties of acreages, or other units of area. However, using three bushels per acre as a rough and generous estimate of seed rate, an idea of yields per acre can be obtained. For wheat, these come out as set out in Table 4.

It must be emphasised that these are only the most general indices of changes in land productivity. Nevertheless, they are consistent with other information. For example, 6 to 7 cwt per acre (750 to 900 kg per hectare) is Holliday's (1962) figure for cereal yields in India a decade ago, an agriculture comparable in status to that of England in the Middle Ages. And it is known that by 1885, English wheat yields were around 18 and 19 cwt per acre (2200 to 2400 kg per hectare). So the trend is right, and provides a pattern against which to consider the husbandry changes that made the changes in yield possible.

In the Middle Ages, a modest level of fertility was maintained by the fallow period in the old open field system of farming, whereby livestock grazed both on the wastes – forest and scrub – and on the fallows, and with the natural practice of bringing them near home at nights, much of the benefit of grazing the wastes accrued to the fallow either as manure from cattle pens, or from night folding on the fallows.

Such practices are to be seen today in Africa, where in Nigeria Fulani herdsmen are paid to kraal their cattle on fallows at night. They illustrate one of the great principles of husbandry, practised by accident or design the world over. This is the principle of fertility transfer. Van Bath discusses it at length in his account of the development of advanced husbandry and high farm productivity in the Netherlands in the 16th and 17th centuries. There a dense population created a demand for both food crops and industrial crops, and also for fuel. In general the farmers on the town margins in Flanders and Brabant concentrated on the high value crops – oilseeds, dye plants, fibres, hops and tobacco and vegetables – and the deficiency in home production of cereals was made good by imports from the Baltic. The main fuel source was peat from the northern provinces of the Netherlands, and peat ash contributed also to the manurial value of town wastes. High value crops, which were generally labour intensive, justified high inputs, and a flourishing trade in manure developed, livestock farms providing farmyard manure, and the towns night soil, refuse, and hearth ash. Thus the fertility of the outlying areas – the Baltic, the coastal stock farms and the peat lands – was transferred to the lands with the better market opportunities, near the towns. These were not necessarily on intrinsically good soils. Van Bath records that the

English traveller, Sir Richard Weston, 'was appalled at the "barrenness of the soil"'.

Fertility transfer, which made high farming in Flanders possible, is very old and very widely spread. As soon as a family sets up a homestead and cooks and feeds continuously in one place, the elements of fertility are accumulated there. There are no records of the places where the first farmers planted their crops, but it is a fair guess that they first used the self-sown plants on their own middens, and they developed their fields in the immediate vicinity of their homesteads, where night soil and refuse would accumulate. The results of this soil amelioration are to be seen in many parts of the world. In India, the land around the old established villages of Malwa varies in quality according to the distance from the village. That close at hand, known as the *adhan* land, has benefited from perhaps 2000 years of casual disposal of wastes and excrement, and is greatly superior both in soil texture and in productivity to the distant *jungle* lands, from which over the same period everything has been removed and taken to the village. In Britain there is a proverb which says 'the nearer the church the better the land'. The church was the centre of the mediaeval village, and though the quality of the land may have influenced the siting of the village, other factors, particularly water supply, were at least equally important, and the chief reason for the superiority of the land near the church is long continued fertility transfer. Even within the village perimeter, the evidence of fertility transfer can be seen in the superiority in soil structure and fertility of the old gardens, both manor and cottage, over the fields beyond the garden fence. In a more recent agricultural context the same process can be seen in the fertility of the household banana gardens of Buganda in Uganda, where productivity is maintained beyond the normal arable cropping period by mulching with household refuse.

Fertility transfer is in a sense robbing Peter to pay Paul. Taking the system as a whole, nothing new has been added. Its importance in the history of agriculture is that it has provided a system whereby high production husbandry could be developed in some favourable situations. It is acceptable since it does not in general result in any irreversible damage to the soils that are robbed. If there are reserves of weathering minerals in the soil, phosphate and potash losses can be made good over time, and losses of nitrogen are in any case made good by the natural processes of nitrogen fixation. So there remains the possibility of spreading high husbandry over the poorer lands as the need for farm produce increases, provided that agricultural technology improves.

The classical example of this spread of high husbandry over a farming area is the 'New Husbandry' developed in eastern England in the latter part of the 18th century. This development is associated particularly with the names of 'Turnip' Townshend and Coke of Holkham, but they were only the leaders of a movement that spread widely in the drier parts of England, and largely on the lighter soils. The inspiration for this English agricultural revolution came partly from the classical education of the English gentry, and partly from their continental travels, where they saw the newly developed high husbandry of the Low Countries so well described by van Bath. From the classical writers they developed ideas on manuring, known and practised by the Romans, but having little part in the low level farming of manorial England. From Flanders they brought new crops that intensified and diversified their farming system, and a concept of fertility improvement that changed their ideas of the potential of their own poor light soils.

The 'New Husbandry' was all that its name implies. It involved an intensification of cropping, replacing the fallows with clover and turnips, and it provided for the full integration into the farming system of the livestock formerly ranging between the wastes and the fallow lands. The addition of clover and turnips to the traditional cereal crops gave rise to the Norfolk four-course rotation, a characteristic feature of the New Husbandry. The new crops made possible the adequate feeding of livestock throughout the year, in itself a substantial intensification of husbandry. This was not all. With adequate feeding, breed improvement became possible, and the superiority of British livestock breeds is founded directly on the improvement in livestock nutrition that was an integral part of the New Husbandry.

Folding of sheep on clover in the summer and on turnips in the winter was found to improve substantially the yield of the subsequent cereal crop. For cattle, buildings were provided for winter housing and feeding. These changes in livestock management led to the change that is the real heart of the New Husbandry. The enclosures had made the farm a unit as it had never been before. The four-course rotation took the sheep round the farm and brought to an end fertility transfer in respect of the sheep flock. The winter feeding of cattle in yards made possible the conservation of all wastes in the manure accumulated in the yards and returned to the land in the autumn. Consider the distribution of cattle yards and barns on English farms. They were planned under the influence of the improvers so that there was a yard and a barn within horse and cart range of every field, thereby making possible the uniform distribution of manure over the farm. Consider also the old tenancy agreements. No tenant was allowed to sell anything off the farm except cereals, meat and dairy produce, thereby minimising the fertility drain, and making effective the policy of conserving the elements of fertility and distributing them, not just on the fields nearest home, but evenly over the farm as a whole. Thus fertility transfer, haphazard and inefficient as it was, gave place to fertility conservation, and to the proper care and improvement of the whole farm, and not just the land nearest the homestead.

Fertility improvement was also undertaken. Lime and marl were used. By the mid 19th century experiments with bones and bone meal had begun. Town wastes were used where they were accessible. Most significant, the nitrogen situation was improved by the regular place of leguminous crops in the rotation. A century of these practices of conservation and improvement over the farm as a whole brought the yields of arable crops to levels never previously reached, and put the soils of England in a condition to respond to the fertilisers that came into extensive use in the mid 20th century.

This was a peculiarly British achievement. The establishment of high husbandry is difficult where pressure on the land is such that there are not the resources available to farmers to capitalise the intensification of their farming. On the other hand, it has no appeal where land is still in excess of demand. Moreover this kind of agricultural revolution has been overtaken by the revolution in agricultural industry, and before considering the prospects of dissemination of high husbandry where agriculture is still at a low level, it is necessary to trace the course of the 20th century revolution in the areas where high husbandry was already practised.

TABLE 5. *Trends in yields of wheat, barley and sugar-beet in Eastern Counties of England: five year averages in quintals per hectare*

Quinquennium	Wheat	Barley	Sugar-beet
1933–7	23.8	21.7	197
1935–9	24.1	22.2	199
1946–50	24.8	23.4	214
1955–9	35.2	29.5	255
1960–64	43.4	33.8	366
1965–9	40.6	37.4	306

From the *Farm Management Survey*, Cambridge University.

The expansion of knowledge on which agricultural improvement is based, is an evolutionary rather than a revolutionary process. Nevertheless, the application of knowledge has gone on discontinuously, and the enormous advances made since 1950 by the application of knowledge, much of which had long been available, has all the attributes of a revolution. In 1939, there arose in Britain a demand for maximum food production that continued for twenty years. During the war and immediate post-war period, husbandry could only be improved within the strict limits of availability of inputs. By 1950 both fertilisers and equipment had become freely available, and the new chemical pesticides and herbicides had been developed. The static nature of British husbandry up to 1950, and the great leap forward that resulted from the application of these industrial inputs thereafter, are illustrated by yield data for three crops in the Eastern Counties of England in Table 5. This rate of increase continued through the early 1960s but in the late 1960s it has slowed down. It was not to be expected that husbandry could be improved so fast for a long time, particularly in a revolutionary situation in which much of the change consisted in making good the lack of progress of most of a century of agricultural depression.

The revolution has been essentially a fertility revolution. It has become possible to make good deficiencies in nutrient elements cheaply and effectively from the fertiliser bag. Natural imbalances have been corrected, as for example phosphate deficiencies on Midland clay soils and minor element deficiencies on fen peats. More than this, new fertility levels can be attained, such as are found very rarely or not at all in natural soils. These high fertility levels have made it necessary to breed new varieties of crop plants, and new races of livestock adapted to high level production in response to high fertility. Moreover, weed and pest control have become more important, since so much more is at risk, and soil management under the new conditions has become more critical, and has called for the speed and efficiency of cultivation and the precision of application of chemicals that only a high degree of mechanisation can give.

Modern industrial inputs have on the one hand released the farmer from some of the limiting factors of earlier farming systems, and on the other saddled him with a need for capital resources far beyond those of his forebears. The chief consequence of this has been a progressive simplification and specialisation in farming practice. Most British farmers now buy their milk from dairies. Very few cure their own bacon. Many do not even keep enough poultry to supply the farmhouse with eggs. Since it is no longer necessary to adhere closely to the old rotations, some farmers have specialised in a few, or even only one crop. Many have either crops only or livestock only, no longer feeling any advantage in mixed farming. Intensification and specialisation to this extent raises its own problems. On livestock farms, effluent and manure disposal is a major problem. On crop farms with some soil types, soil management problems are becoming increasingly serious in a system involving continuous tillage and no organic manure. High husbandry, like other advances in technology, creates more problems than it solves, but in the solution of new problems lies the hope and the challenge of agricultural advance.

This is a human enterprise. Even the most mechanised agriculture still needs farmers, though too often new techniques and new opportunities are discussed without reference to the men who farm, and on whose capacity for change farming advance depends. It is indeed the greatest achievement of British farming that British farmers have proved sufficiently enterprising and receptive to new ideas, to have made these vast changes, and to have acquired the management skill to organise their farming on a scale and at a pace not hitherto contemplated.

The pattern of intensification of husbandry, first by the conservation of the elements of fertility, and later by the addition of these elements by the use of fertilisers, is peculiar to Britain and to some neighbouring countries of continental Europe. Elsewhere, fertiliser practice has been established on soils at a low level of productivity, either as a consequence of long ages of exploitive agriculture, as in India, or through the intrinsic poverty, in terms of plant nutrients, of the soil forming material, as in Australia.

India is endowed with vast tracts of soils that are intrinsically fertile. The low level of production of Indian agriculture until recently represents a basic fertility level to which the soils have been reduced by upwards of four millennia of exploitive agriculture. That they still produce enough to maintain the dense population of the Indian sub-continent is sufficient evidence of their enduring fertility. Indian agriculture has other assets also. Monsoon rains, though generally regarded as uncertain, are adequate and reliable compared with, for example, the rainfall pattern in Australia. The degradation of the soils has come about because of the limited power (oxen and buffaloes) available for cultivation, lack of conservation and distribution of manure, and pressure of population on the land, leading among other things to the consumption of the greater part of the manure as fuel.

Escape from this cycle of poverty has come from the use of fertilisers to raise fertility levels, and not, as in Britain, from the conservation of the elements of fertility already there. The careful collection and distribution of organic wastes as a means of raising fertility levels was advocated by Howard in the 1930s, but it never became an important factor in Indian farming. On the other hand, fertiliser use together with better use of water on irrigable land, followed by the introduction of new wheat varieties, have led to new standards of husbandry that are described as a 'green revolution'.

By contrast, in Australia soils of even modest intrinsic

fertility are rare. Most soils are extremely poor, and water resources are scanty and uncertain. Yield levels, either of crops or of forage, are low, and production depends on the exploitation of vast areas at a very low human population density. The pastoral sector is dominant. Merino sheep for wool, and cattle for beef and for dairy produce have been the mainstay of development.

The poverty of Australian soils has meant that intensification of husbandry has depended heavily on research on soil fertility. The identification of nutrient deficiencies, particularly minor element deficiencies, has opened up the prospect of fair returns from formerly uncultivable land by comparatively small investment in fertilisers. With large resources in land, with the prospect of improvements in productivity by fertiliser application, and with the capital resources of a country geared to the Western economic system, intensification of husbandry has been rapid and effective. This agricultural pattern will continue. The pastoral areas will remain pastoral, and crop production will be limited to present areas and such others as can be brought under irrigation, because of the overriding limitation of water supplies. Nevertheless, further substantial changes are probable. The prospect of increasing productivity on the better lands is such that Australia will be able to produce all the agricultural goods for which there is a market without using the marginal lands of the continent. With the population at levels that can at present be foreseen, some withdrawal from the poorer and more difficult lands now in agricultural use will be possible, and indeed politic.

In India it did not prove possible to base the improvement of husbandry on fertility conservation. In Australia there is little or no fertility to conserve. Nevertheless, the matter does not end with the substitution of industrial fertilisers for conservative farming practice. Recent experience in Britain has given force to the argument that the management of soil fertility involves more than the chemistry of the elements of fertility. The biology and the physical structure of soils are equally important in productivity. In Britain it has become necessary to look again at the means whereby the soils of the country gained in fertility under the New Husbandry. In Australia it will be important to study means whereby these components of productivity can be

53

established for the first time. In India, with the great range of soils and husbandry circumstances, there is the experimental material awaiting study on which to base a theory that will integrate soil biology and soil physics with nutrient chemistry and water management. For all these together comprise the science of husbandry.

CHAPTER 5

TECHNOLOGY

Agriculture began as a craft, but its development has been furthered by technology from very early times. Among the first applications was the use of mathematics, astronomy, and engineering in the provision of water for irrigation in desert countries. Mathematics and astronomy were of importance in the prediction of seasonal river floods, and hence were developed in Iraq, northern India and Egypt. Prediction of the flood was the first step. As population grew and the demand for food increased, control of the flood and the distribution of water became necessary. So engineering was required. Simple flood exploitation, still to be seen in the deltas of the ephemeral rivers of the Arabian peninsula and the Red Sea coast in the Sudan, gave place to well managed irrigation schemes such as those on the rivers of Egypt, northern India and Pakistan, and on the ancient tank systems of Ceylon. These are only possible where there is a well developed social structure, with a strong and stable administration. Under these circumstances the necessary technology can be generated, and the increases in agricultural production that can be achieved are very great. The agricultural productivity of Egypt, West Pakistan and Northern India all depend on great engineering works that could only have been undertaken by well organised, socially advanced communities. Such communities in their very nature have a powerful political sector.

The application of technology to agriculture was greatly accelerated by the Industrial Revolution, and the pattern of modern agriculture is very largely the consequence of the demands on agriculture of an increasingly urban society, and the technological inputs that urban society makes possible. First and most important was the industrial impetus to the development of the American West. The British market for cereals, and later for meat, the heavy industry able to provide materials for ship and railway building, and the research resources for the study of food preservation, were responsible for the opening up of America, Australia and New Zealand for agriculture, and for the subsequent vast flow of food for Britain along the rail and steamship arteries thereby created. It is not often appreciated that an early stage in the mechanisation of agriculture is the provision of transport. In Britain the stimulus of cheap and rapid transport was of enormous importance in agricultural development. The railway brought the cheese producing areas – Cheddar, Cheshire, Stilton – into the liquid milk market, and that innovation was followed later by the road tanker and the bulk collection service. Local markets and local prices gave way to a country-wide marketing system, and a price structure dominated by the demands of the great cities.

The communications revolution brought a much larger sector of agriculture within range of the big and growing consuming centres. Rail and steamship moved the cereals and meat of North America and the Argentine. Head-load, river and lake steamer, rail and ocean liner were necessary to move the cotton and later the coffee of Uganda. Head-load gave place to bicycle load and then to lorry, and it was on lorries transporting cotton and coffee that Uganda first learned to drive internal combustion engined vehicles. Transport has made possible specialisation between countries. Without it there would be no cereal mono-culture in North America. On it depend also the world's great plantation crops, which involve the provision of tropical produce for temperate markets. It took capital, technology and manage-ment from industrial countries, with labour and vacant land in developing countries, to grow the produce for which there was a market in the industrial countries.

The impact of the industrial revolution on agriculture was soon felt in the East. Cotton goods first came to Britain from India and were for long discouraged by tariffs or prohibitions to safeguard England's wool. The invention of textile machinery and the development of the cotton crop in the American colonies led to a reversal of the flow of trade. India, formerly an exporter, became an importer on a massive scale. Indigenous manufac-tures were replaced, and her cotton crop deteriorated in quality to the level at which it could be maintained by low grade exports to China, and later to Japan. Only by the estab-lishment of cotton mills in India and the imposition of a tariff on Lancashire's exports to India was India able to regain her place in the textile world, and to re-establish good cotton varieties in her agriculture.

Western technological influence has been felt in other ways also in agriculture. The development of cattle ranching in America (north and south) called for such industrial inputs (besides transport) as fencing materials and meat factories. Indeed, the technology of the Industrial Revolution, which began in England, has had a stimulating, and sometimes devastating, effect on agriculture the world over, because of its enormous impact on transportation and on the generation of effective demand. Thus it came about that after the initial stimulus to British agriculture (up to about 1860/1870), there ensued a long period of agricultural depression. The agricultural depression was not confined to Britain. North American farmers suffered also. Industrial inputs into agriculture had brought in new lands at a rate that led to overproduction in all those sectors of the world where there was a money market for food. This situation changed in 1940. For 25 years food needs exceeded food production, and the stimulus of the need has led to a fresh surge in the application of technology to agriculture.

This most recent contribution of technology to agriculture has been made in three categories: power and machinery, fertilisers, and chemical pesticides and weedkillers. The development of farm machinery has gone on over a long period, and in England the 19th century saw a great and progressive development of tools of cultivation and harvesting. Power came in with the use of steam for threshing, and at the end of the century for ploughing and cultivation. But it was not until the 1914–18 war that power from the internal combustion engine came on to British farms. Even then, it was said that though the tractor was a useful supplement to the horse, it could never economically replace it. The horse equipment and layout was designed for horse efficiency, and the tractor only really became established when the power lift, implements designed for tractors, and rubber tyres were added to it. Now, of course, the horse has so declined in Britain that it is no longer worthwhile collecting statistics of farm horses. Modern tractor power, developed during and since the last war, has revolutionised British farming, in that farm operations can be more timely carried out, more effectively done, and done with very much less labour.

Mechanisation, and particularly the introduction of the tractor, has been popular throughout the developing world. It has featured extensively in aid programmes, and in the 1950s in particular there were very numerous projects for the 'modernisation' of agriculture by the introduction of power and equipment from the industrial to the developing countries. Experts and advisers as well as recipients in developing countries learnt by experience, and at great cost, that mechanised agriculture is not established in a country just by the import of equipment and the training of tractor drivers. The understanding of mechanical equipment, and the appreciation of its capacity and its limitations that is almost second nature to the British farm worker, is not acquired by his counterpart in a developing country in a short mechanics training course. Maintenance is expensive and, at least initially, inefficient. Supplies of spares can rarely be maintained at an adequate level. More important than all these disabilities, modern mechanisation is designed and developed to economise in labour where labour is expensive and has a wide range of alternative sources of employment. It has little relevance where labour is the only abundant resource.

In the 1960s a more balanced appraisal has emerged of the place of mechanisation in development. It is generally recognised that there is nothing to be gained by substituting machines for labour where no alternative employment exists. The pressure for rapid 'modernisation' has gone, and in its place is a more realistic approach to the use of power and equipment for such work as dry weather cultivations that could not be done at all with manual labour or oxdrawn implements, and to meet peak demand, such as wheat threshing in north western India, for which the labour supply is insufficient.

The fertiliser industry is the greatest contribution of technology to agriculture. It is derived from the enquiries of chemists and agriculturists of the 19th century into the nature of soil fertility. There followed the experimental production of mineral substances for use as fertilisers, culminating in the fixation of atmospheric nitrogen at the beginning of the 20th century. On this experimental work the whole of modern fertiliser practice depends. It is thus a very recent development. In 1845 the United Kingdom used 30000 tons of nitrogen fertiliser per annum, all guano. And the use of other fertilisers was equally negligible. Up to 1916 the total amount increased only very slowly, though guano gave place to sodium nitrate and ammonium sulphate. Consumption of nutrients ($N + P_2O_5 + K_2O$) only rose from 0.05 million tons in 1845 to 0.23 million tons in 1916, and to 0.25 million tons in 1942. From then it went up steadily to nearly 1.5 million tons in 1964.

The fertiliser revolution has had a very different impact in different parts of the world. Overall, world fertiliser consumption has gone up from 2 million tons of nutrients ($N + P_2O_5 + K_2O$) in 1906 to 29 million tons in 1962. Much the greater part of this has been in the advanced agricultural systems of Western Europe and North America. Special needs have dictated special policies. In Australia, for instance, the ability of legumes to fix nitrogen has been exploited by a policy of meeting soil needs for other nutrients, particularly phosphates and minor elements, and thereby encouraging good performance in leguminous forage crops as a means of providing the necessary nitrogen without industrial imports. Introducing fertiliser use in Indian agriculture was more difficult. In an ancient agriculture, carried on by very poor peasant cultivators, a new practice involving substantial outlay at the beginning of the cropping season was hard to establish. A beginning was made with nitrogen, being the element in shortest supply. The Australian approach was not possible, as pressure on the land is too great for the large scale planting of leguminous forage crops. Industrial nitrogen was therefore used. It was followed by phosphatic fertilisers, and by other elements where required.

Once established, fertiliser use grew rapidly, and in recent years the introduction of new fertiliser responsive crop varieties has greatly accelerated the rate of growth of fertiliser consumption. India is beset by the problems of shortage of capital and of foreign exchange. The policy in building up fertiliser use has been affected by this broader economic situation. For not only does the farmer need capital to buy fertiliser from which he will only reap a return when the crop ripens, but the industrialist needs both capital and foreign exchange to set up a fertiliser factory. India's fertiliser industry has grown rapidly, but the fertiliser demand that is an integral part of the current 'green revolution' has grown much more rapidly. So both home production and imports have increased, and are expected to increase

TABLE 6. *Fertilisers in India: home production and imports: thousand tons*

Year	N Production	N Imports	P_2O_5 Production	P_2O_5 Imports	K_2O Production	K_2O Imports
1952–3	53	44	7	–	–	3
1957–8	81	110	26	–	–	13
1961–2	154	143	65	1	–	30
1965–6	238	376	119	22	–	94

further in the next planning period. The growth of India's fertiliser production and fertiliser use are to be seen in the figures in Table 6.

In Africa, soil fertility is lower than in India, but population pressure is low and unused land is still available. Production is more easily increased by extending the cultivated area than by fertiliser practice. Hence the fertiliser revolution has still to come. In particular areas where land values are high and production can be increased thereby, fertilisers are already in use. The capital involved in establishing irrigation systems such as those of the Sudan, and plantation crops such as tea and sugar, is such that it is worthwhile to maximise yield per acre by fertiliser use. Meanwhile, research on soil fertility and fertiliser response is building up a body of knowledge on which rational fertility improvement can be based in due course.

The third great class of industrial goods that is used in agriculture is the group of chemicals on which modern weed and pest control is based. This is the most recent contribution of technology to agriculture. Indeed the invention of the chemicals in this class scarcely goes back before 1945. In Western Europe and America they are very widely used indeed, and the dominance of the farmer over weeds and pests has been enormously increased. Application in the rest of the world has begun, and has had striking effects on some crops, for instance cotton, but costs in terms of foreign exchange, and technological and educational limitations to the use of high precision techniques, are such that spread of use of these chemicals will not be as rapid as of fertilisers. They are powerful and often persistent poisons. Their misuse has had serious side effects on the natural fauna and flora, some of which have reacted unexpectedly on the agricultural system. The problems they create must be solved by rigorous testing and control, and by intelligent and responsible use.

The magnitude of the changes that have resulted from the application of the new technology to agriculture in Britain is indicated by the fact that output has doubled in the last quarter of a century. Its effect has been through new thinking, and the application of new systems to agriculture. The introduction of the new resources was of little agricultural value until the agricultural system was changed. Thus in Britain, tractors replaced horses where the whole approach to cultivations was modified, and fertiliser practice, weed control, and pest control form part of an integrated system of high production. This in turn is only successful because plant breeders have produced a range of varieties that perform successfully under these high farming conditions. Similarly, in African countries the production of the cotton crop with the use of modern insecticides is only profitable when agronomic practice is such that plants capable of carrying a large crop are regularly grown. These conditions apply equally to livestock. Reference has been made to Joblin's demonstration of the low milk production ceiling of Teso cattle. Preventive veterinary medicine is the first step in improvement and adequate feed the next, but there will be no adequate return for drugs and sprays until better nutrition is matched with better breeds.

CHAPTER 6

NUTRITION

No discussion of farming and food supply would be complete without some consideration of nutrition. Quality and balance in the diet are second only to adequacy in amount of food available. Moreover, agricultural policy has been greatly influenced by the advances in recent years in the specification of human dietary needs.

Knowledge of the basic requirements for good dietary balance began with Hopkins' identification of vitamins in the years immediately after the first world war. Progressively the nature and role of these accessory food factors, vital for health but only required in very small quantities, have been brought to light, and their natural sources have been identified. After the second world war, research on diseases of children in developing countries revealed a further major cause of ill health of dietary origin. This is the inadequacy of the protein content of the food supply. Since the requirement for protein is substantially greater during growth than at maturity, protein malnutrition is particularly severe in children. The protein deficiency disease of young children has been identified and described, and is now well known under the name of 'kwashiorkor'. More recently the long known difference in nutritive quality between protein of animal and vegetable origin has been analysed and quantified in terms of amino acid composition.

As a result of these great advances in nutrition studies, the agriculturist now has a specification of the food requirements necessary for good health in man. First comes adequacy in quantity, enough food to meet the daily calorie requirements of the body. Next is adequacy in protein to meet the tissue building and repair processes of the body. This is not a simple requirement, since needs change with age. Quality must be such as to provide the body at each stage in growth with the amino acids necessary for the synthesis of body proteins. Third is the requirement for vitamins, and for certain minerals that are also essential in small amounts. Evidently all these needs have been met over most of history and in most areas of human habitation, or the human race would not have survived. The problem is to identify those areas where diet is deficient, and to study the reasons for the deficiency and the means to make it good.

The two root causes of malnutrition are ignorance and poverty. Given a wide range of choice, man like most other animals, would naturally achieve for himself a sound food balance. But with increasing populations he has pressed on his food supply in two ways. First, he has spread into areas where climatic conditions have limited his agricultural activities, and second, he has developed large populations of whom a substantial proportion are too poor to exercise any choice in their diet.

Consider two examples of spread. In the Caribbean and on the Spanish Main, the native American roots, cassava or manioc and sweet potatoes, produce ample quantities of starchy calorie supplying food. They are adequately balanced where supplies of fish are available from the sea, or the great rivers. Inland in forest clearings, manioc does admirably but the protein supplement is not available, and the quality of the diet is consequently poor. Similarly, in the well watered regions of equatorial Africa, a series of communities has arisen dependent primarily on green bananas as their calorie source. Until recently these regions supported large game populations, and while these persisted there was available an adequate source of high value protein to supplement the starchy food. With the recent increase in population the game was overwhelmed, protein supplies fell and diet deteriorated.

Diet in the densely populated Asian countries has suffered through poverty. Grain and starchy roots give the greatest return to labour and to land of all crops and livestock. Where population pressure builds up, food for livestock gives place to food for people, and vegetable crops and protein crops give place to cereals and root crops. Hence in India profit maximisation tends to eliminate pulse crops, simply because poor people go for the cheapest food, and consequently the price differential between grains and pulses is not great enough to counterbalance the yield differential. That this trend is real, and that it has not been halted by the growing appreciation of nutritional needs in recent years, is shown by the statistics of Indian food grain production. In 1950–51 pulses comprised 16 per cent of all food grains. From 1960–61 to 1964–5 they were about 14 per cent, and by 1968–9, with the impact of the high yielding wheat varieties on the cropping system, they had fallen to 11 per cent. The same problems beset the growing towns of Africa. Poor urban labourers buy cassava, sweet potatoes, and green bananas, all with a very low protein content, and cannot afford meat or fish or even beans, except rarely for a feast. Since in neither type of malnutrition circumstance do the people directly affected know the elements of nutrition, there is no incentive for amendment.

In all countries where agriculture can be practised, a balanced diet can be produced, but a good diet costs more than a poor one in land, labour, knowledge and human skills. Calorie needs are the most easily satisfied. Grains, roots and tubers produce 'fuel' for the body in quantity, and other foods such as proteins are used to supply calorie needs, if necessary. Until recently there was great anxiety lest it should prove impossible to meet the calorie needs of the human population. The great agricultural advance in India in the last few years has demonstrated the existence of an agricultural potential sufficient in the short term to meet the needs of the growing population.

Protein needs are more difficult to meet. Protein is complex, and of its component amino acids not all can be manufactured in the human body. Animal protein gives the most suitable amino acid balance for human needs, and is widely used where available. Vegetable protein is a natural component of all vegetable materials, but the proportion in which it is present varies widely. The world's major vegetable foods may be classified roughly into three groups. Roots and tubers and other underground storage organs, together with the food bananas, have the lowest protein content, of the order of one or two per cent. Cereals range from 10 to 15 per cent. Leguminous seeds (pulses) average about 20 per cent. Evidently a diet of roots or tubers, or food bananas, must be supplemented with other protein to provide an adequate diet, even for an adult. The better cereals, on the other hand, come close to meeting the protein requirements of adults, but are inadequate for growing children. The importance of the pulses, with a high enough protein content to supplement either cereals or root crops, is readily apparent. Where animal protein is not available in sufficient quantity, the provision of a balanced diet must depend heavily on the provision of suitable pulses.

Advances by breeding in pulse crops have not yet matched those achieved in cereals. The importance of pulse crop improvement should be emphasised. Yield improvement is a primary consideration, and in the range of material now available scope for improvement is apparent. Substantial advance in production per day has been demonstrated in short term races of *Cajanus* and *Phaseolus* bred in India. Prospects of higher yield in the American species *Phaseolus vulgaris* are apparent in African breeding projects. Moreover, quality characters are also variable in pulse species, and improvement by breeding may be expected.

Vitamins and minerals are obtained from a wide range of vegetables. By the exploitation of the known range of species and varieties, adequate dietary supplements can be provided wherever agriculture is practised. As with proteins, so with accessory food factors, dietary balance is a matter of knowledge and of farming resources.

By the application of science and technology, high farming capable of supplying a generous and balanced diet to vast urban populations has been established in the western world. Enough research and development has gone on to show that the agriculture of the rest of the world could be similarly improved. It must be the task of the next half century to achieve this improvement if the human populations that can already be foreseen are to be adequately fed.

PART III: SYSTEMS OF AGRICULTURE

CHAPTER 7

BRITAIN

Modern British agriculture, and British based agricultural systems in other parts of the world, depend upon the close integration of agricultural and industrial production. This close integration mediates the exchange of food from agriculture for industrial inputs – power and machinery, fertilisers, and agricultural chemicals – from the urban sector and has been the means whereby agricultural productivity has been raised enormously while the human resources devoted to agriculture have steadily declined. The development of this relationship has gone on over the past two centuries, through three distinct phases. First was a phase of developing industrial strength supported by an agriculture increasing in efficiency and contributing substantially by the release of labour to industry. Agriculture grew in prosperity as it met the growing need for food in the expanding industrial cities. This was followed by a long period of depression in agriculture when the export of industrial goods had led to the opening up for farming of vast areas of empty land. Agricultural production outstripped urban demand and the food markets of the world were oversupplied. In the third period the absorption of world food surpluses in the 1939–45 war, followed by the achievements of preventive medicine in the post-war period in increasing the rate of population growth, led to 25 years of heavy demand for agricultural produce.

This chapter will be devoted to the pattern of agriculture in the third period, established in Britain during the last war and developed in the two following decades of strong demand. Britain entered the last war with her agriculture suffering from the deep depression of the second period, with much land out of cultivation and that which was being farmed predominantly in

67

grass. Wartime policy was to increase as far as possible the production of food for direct human consumption, which meant producing the maximum of cereals, potatoes and sugar-beet at the expense of livestock. Pigs and poultry were particularly heavily reduced, as much of what they eat can be made suitable for human consumption. Ploughing up grassland for

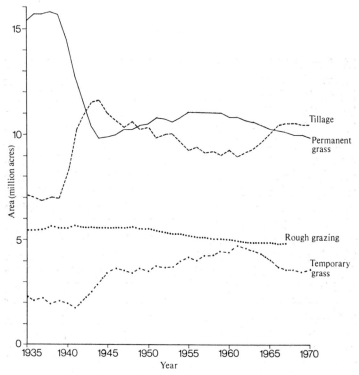

Fig. 1. England and Wales: use of agricultural land.

cereal production made it necessary to reduce grazing livestock, and sheep and beef cattle were cut down. Dairy stock were kept, as a matter of policy, and it was thus possible through milk production to maintain a good nutritional balance in spite of the reduction in meat and eggs. The extent of the change in British farming is best indicated by the change in the proportions of the land devoted to grass and to tillage. Data for England and Wales (abstracted from the Ministry of Agriculture annual returns, HMSO) are set out in Fig. 1. Between 1939 and 1944 the acreage of permanent grass fell from $15\frac{1}{2}$ million acres to $9\frac{1}{2}$ million acres, and the tillage acreage went up from under 7 million to $11\frac{1}{2}$ million.

After the war, in spite of the continuing need for high cereal production, the tillage acreage fell slowly until, by 1960, $2\frac{1}{2}$ million acres had been lost to permanent and temporary grass. Thereafter, it rose again, and by 1967 had reached $10\frac{1}{2}$ million acres. The ploughing up campaign involved more than the simple replacement of permanent grass by tillage, and the husbandry factors involved are reflected in the change in the acreage of temporary grass. The wartime decline in permanent grass was followed by a rise in temporary grass, reflecting the need that could not be long postponed for a grass break. In the decade of the 1950s, the temporary grass acreage rose steadily at the expense of tillage, while permanent grass was little changed. In the 1960s, the increase in tillage was met in part by a fresh reduction in permanent grass, but with new found confidence in the ability to farm without grass and the accompanying livestock, the grass break represented by temporary grass was also reduced. The tillage acreage has not increased since 1967, and it may be that this reflects the anxieties that have developed since about 1955 as to our ability to maintain good conditions, particularly on clay soils, without the break from tillage that can only be obtained by a return to grass.

Rough grazing acreages are also plotted. They show a small but steady decline. Some of this may be attributed to improvement and transfer to permanent grass. Some is due to retirement from agriculture, primarily for forestry but to some extent for recreation and for water gathering.

Changes in livestock populations in England and Wales are plotted in Fig. 2. The heavy wartime reduction in poultry, pigs and sheep is in sharp contrast to the slow increase in cattle numbers in consequence of the support for milk production. Between 1939 and 1944 poultry, pig and sheep populations were reduced to less than half, whereas the cattle population went up by $7\frac{1}{2}$ per cent. It will be seen that the poultry population fell to a minimum in 1943 and thereafter increased rapidly. An incipient recovery after 1943 to 1944 is also to be seen in the data for pigs and sheep, but there followed a further fall,

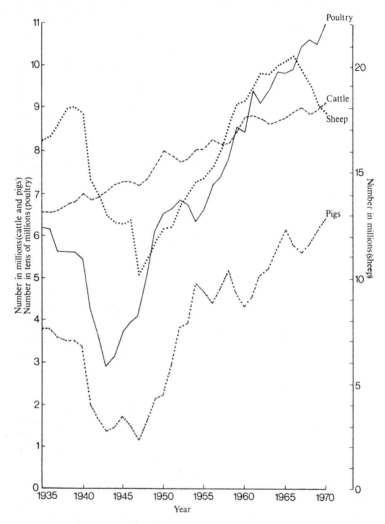

Fig. 2. England and Wales: livestock numbers.

catastrophic in sheep, and serious in pigs, in 1947. This was the effect of the very severe winter on a farming system that had not recouped its reserves after the restrictions of the war period. It will be seen that the severe winter of 1962–3 had no such disastrous consequences.

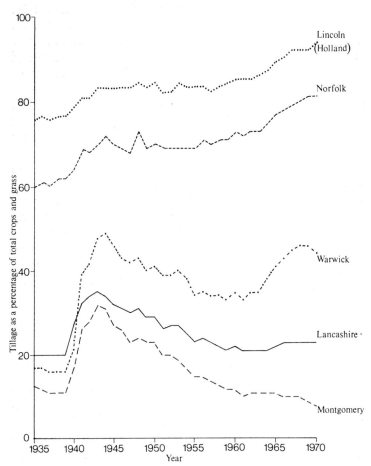

Fig. 3. England and Wales: tillage as a percentage of total crops and grass in five sample counties.

The overall change in the pattern of use of farm land in England and Wales is the sum of different land use patterns in different parts of the country. These detailed patterns are greatly influenced by the trend of decreasing rainfall from west to east. To illustrate the more important differences, data for tillage intensity and livestock populations are plotted in Figs. 3–7 for five counties (Norfolk, Lincoln (Holland), Warwick, Lancashire and Montgomery) across the breadth of mid-

England and Wales. To facilitate comparison between counties that differ in size, data are expressed as acreages, and as stock numbers, per 100 acres of crops and grass. In comparisons of stocking pattern it seemed desirable also to give an index of stocking density for grazing livestock, and for sheep and cattle a second chart is given, showing number of head per 100 acres of all grazing (permanent and temporary grass plus rough grazing).

Data for tillage intensity are given in Fig. 3. The low rainfall areas of eastern England had a high proportion of land under tillage before 1939. The increase during the war was substantial, but not spectacular. There was no fall after the war, and there was a further rise from 1957 onwards. Nearly 95 per cent of the land in the Holland division of Lincolnshire is now under the plough, compared with only 83 per cent from 1943 to 1956, and only 76 per cent before the war. The Midlands, represented by Warwick, on the other hand, entered the war with a small tillage acreage, increased it enormously to meet the wartime need, and turned about half the extra tillage back to grass after the war. Then in the 1960s there was a return to tillage, though not sufficiently to regain the wartime maximum. The west of England and Wales also entered the war with a low tillage acreage and increased it greatly during the war, but the whole of the increased tillage acreage was returned to grass by 1960. The return to tillage that occurred in the Midlands in the 1960s did not take place in the west. Tillage percentage hardly changed in Lancashire, and actually fell further, and well below the pre-war figure, in Montgomery.

Data on cattle populations are given in Fig. 4. Numbers per unit agricultural area (Fig. 4A) are highest in the west and lowest in the east. Moreover, in the west they have steadily increased. In the Midlands (Warwick) they have hardly changed. In the east, Norfolk showed some increase in the 1950s, but in the 1960s numbers fell to the pre-war figure. In Lincoln (Holland) there has been a tendency to decline throughout, considerably accelerated in the 1960s. The county now carries little more than one third of the cattle stock it had before the war. Intensity of stocking expressed as cattle per 100 grazing acres (Fig. 4B) is highest in the east and lowest in the west. In Lincoln (Holland) the figures mean little, as there is little grazing and few cattle. For the other counties, there has been a steady

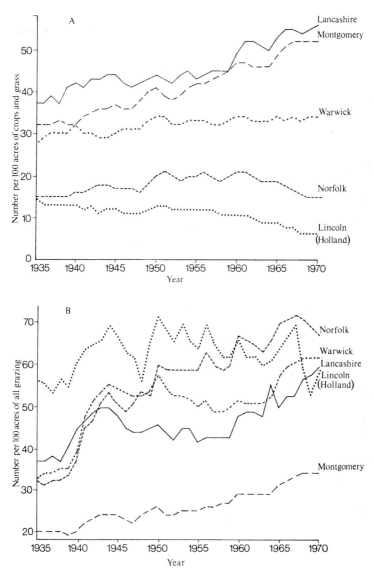

Fig. 4. England and Wales: cattle populations in five
sample counties.

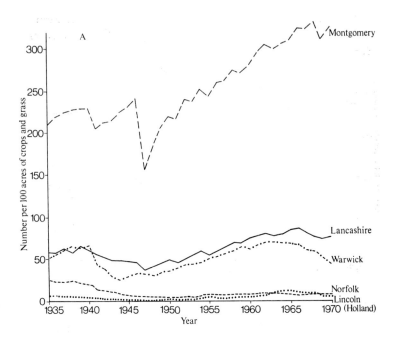

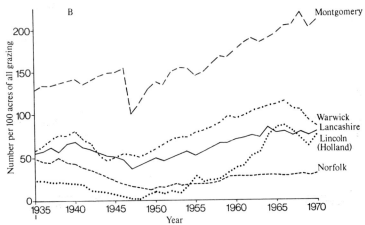

Fig. 5. England and Wales: sheep populations in five
sample counties.

increase in stocking density. In Lancashire and Warwick this was particularly marked during the war, and there was some decline in post-war years. However, in Warwick, Lancashire and Montgomery, the stocking rate for cattle is now higher than it has ever been.

Data on sheep populations are set out in Fig. 5, A and B. Sheep are unimportant in eastern England. In Warwick, following the wartime decline, there was a steady rise until in the early

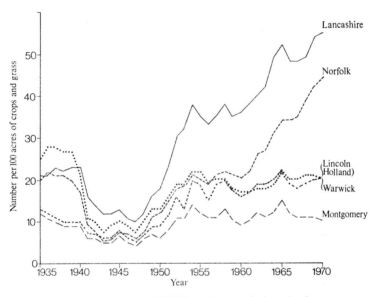

Fig. 6. England and Wales: pig populations in five sample counties.

1960s numbers were as high as pre-war, and stocking rates substantially higher. In the late 1960s both numbers and stocking rates fell. Evidently the falling grass area was devoted to cattle at the expense of sheep. In Lancashire the post-war recovery took sheep numbers well above pre-war, and there has only been a slight fall in the late 1960s. Sheep stocking rates have also been maintained well above the pre-war level, even though the stocking rate for cattle has been steadily rising. Of the five counties, only Montgomery has a large sheep population. The drop during the war was small, as might be expected. The hill grass on which the sheep are kept was not suitable for alternative uses, even under wartime pressures. The big fall in sheep numbers was in the disastrous 1947 winter. Thereafter, numbers and stocking rates increased steadily until 1968.

Before 1939 the largest pig population (Fig. 6) in the five counties was in the Holland division of Lincolnshire. Next came Norfolk and Lancashire, while in Warwick and Montgomery few pigs were kept. Following the heavy wartime drop, Lincoln (Holland) only recovered to about 70 per cent of

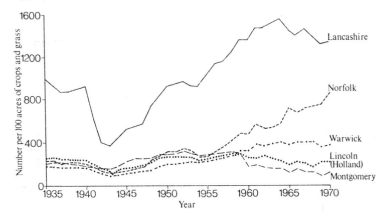

Fig. 7. England and Wales: poultry populations in five sample counties.

previous numbers. Montgomery regained pre-war numbers and Warwick rose by about 60 per cent. In Norfolk and Lancashire, on the other hand, there was a rise (with the fluctuations characteristic of pig populations) until in 1970 both counties had more than twice as many pigs as pre-war.

Poultry keeping has changed more than any other livestock enterprise in recent years, and these changes (Fig. 7) have been associated with a rise in numbers in England and Wales from about 60 million pre-war to 110 million in 1970. In the five counties this change is reflected almost exclusively in the data for Norfolk and Lancashire.

The parallel between pigs and poultry in county pattern is striking. The existence of a vast urban market virtually at the

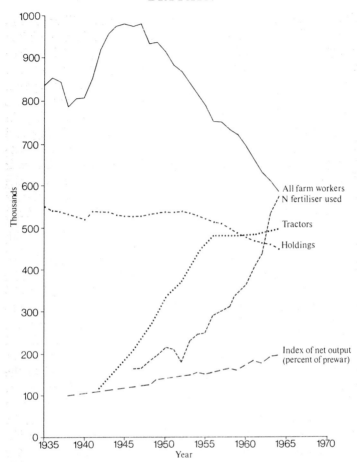

Fig. 8. Changes in resources devoted to agriculture and index of net agricultural output in UK, 1935–64. (From Ministry of Agriculture Statistics and Annual Abstract of Statistics.)

farm gate in Lancashire undoubtedly acts as a strong stimulus to intensive enterprises such as pigs and poultry, and also horticulture. But the stimulus of the proximity of the west Midlands cities has not led to the same response in Warwick, and the development in Norfolk cannot be ascribed to the strength of a local market.

These changes in land use and stock numbers and distribution have gone on in a context of great changes in the resources devoted to agriculture. Some of these changes are illustrated in Fig. 8. The number of agricultural holdings in the United Kingdom remained fairly constant until 1953, and has since slowly declined. Land has been lost from agriculture to urban needs at a rate of about 40 000 acres per annum in recent years, and this together with amalgamation of uneconomic units has resulted in a fall of 14 per cent in the number of holdings in 10 years. The labour force in agriculture rose during and immediately after the war, but has fallen steeply from the post-war peak in 1947. Between 1947 and 1964, the total labour force declined by 39 per cent. Industrial inputs rose as the labour force declined. The number of tractors in 1956 was nearly two and a half times that in 1942, but after 1956 did not increase appreciably. Fertiliser inputs, on the other hand, continued to increase, and the use of nitrogen – the most important fertiliser element – was four times what it had been in 1942, and was still increasing rapidly. The combined effect of the changes in the resources devoted to agriculture is to be seen in the index of net output. This rose from 100 pre-war to 128 in 1948, and 198 in 1964, amounting to doubling in 25 years.

These changes have been brought about by the application of science and technology to agriculture in a period in which the strong demand for farm produce has been transmitted to the industry through the system of State support for farm prices. Considering first the impact of technology on agriculture, there has been a very great advance on each of the main sectors of the application of industry to agriculture. Fertiliser practice has improved out of all knowledge, partly in application on the farm, but equally in formulation in the chemical factory and in presentation by the trade. Modern fertilisers contain a high concentration of nutrient elements, carry very much less of unwanted chemical constituents and are formulated in compounds suited to farm needs, and granulated for ease of distribution. The whole range of chemical pesticides and herbicides is new to agriculture. These are powerful weapons and in learning how to use them, harm as well as good has come both to agriculture and to wild life. Nevertheless they have greatly enhanced agricultural production and, as their use has been mastered, they have become increasingly valuable.

Farm power and machinery has undergone development that

has been equally revolutionary. The diesel engine, the rubber tyre and the power lift have made the tractor an efficient and versatile instrument. Implement design has progressed to take advantage of the new tractors, not only to perform the same cultivations better, but also to meet new needs. In modern farming, precision machinery is now available for the application of sprays in small quantities, seed dressing, fertiliser distribution and for seed sowing.

These industrial inputs into agriculture could not have been absorbed effectively unless the farmer and the farm worker had changed to deal with them. They have become in the course of a single generation one of the most skilled and versatile groups in the British work force. In this evolution of new skills the agricultural education and advisory system has played a large part. The National Agricultural Advisory Service was set up after the last war to rationalise and establish on a permanent footing the advisory services that had been expanded on an *ad hoc* basis to meet a wartime need. They served British agriculture in a wide variety of ways. In the immediate post-war period they provided the advice which ensured the proper application of the new and increased resources that became available, and hence guided the increase in production at a time when food was needed at almost any price. In the latter part of the decade of the 1950s, when food production began to overtake demand, it became necessary to call a halt to the rise in the cost of Exchequer support for agriculture. It then fell to the Advisory Service to move from the field of technological improvement, for which they had been recruited and in which they had had their training and experience, into the field of economics and management, for which they had little previous preparation. It is one of the great achievements of agriculture in the mid-century that the Advisory Service undertook its own retraining, and over a period of a few years established with practising farmers an understanding in the new circumstances that led to as great an advance in British farm management as had previously been achieved in farm technology.* The Advisory Service led the way. The problems pressing on the farmer on the one hand, and arising from the impact of economic circumstances on State

* The reorganisation of the NAAS to form the ADAS took place after this was written.

policy on the other, were their immediate responsibility. Farm institutes, agricultural colleges, and university faculties of agriculture, experienced these pressures later and less urgently. There is a case for a closer integration of the specifically educational institutions with the Advisory Service. Nevertheless, the existing informal association has led to progressive changes in the educational system whereby the gains achieved and the needs felt by the agricultural industry have been taken into account in the curricula of agricultural education. In a time of rapid change this has not been achieved without strains and disagreements, and it is of the greatest importance for the future health and success of agriculture that these should be resolved.

The support system by which the remuneration of the agricultural industry has been determined for a quarter of a century must now be considered. State control of agricultural prices is no new thing. 'Beginning in the early Middle Ages, and ending in 1869, the English Corn Laws lasted for upwards of six centuries. Attention has been so exclusively concentrated on one side only of their provisions, that the regulation of the inland trade in corn and the restrictions on its exportation have been long forgotten. . . . The general aim of legislators was to maintain an abundant supply of food at fair and steady prices . . .' (Ernle, 1919.) During this period, corn prices in Britain were manipulated by export embargoes when prices rose after a short harvest, and by export bounties when they fell in years of abundance. As the population grew, the circumstances in which an export bounty was called for gradually disappeared, and by the mid 19th century, the Corn Laws operated one way only, as a tariff on imports. Yet up to the middle of the 19th century, the growth of the British population was matched by the increase in British agricultural production, and imports and exports of food grains were of marginal significance. By 1811 British agricultural production was feeding 11 million people, and by 1841 it was feeding 16½ million. Through this period there were times of agricultural prosperity and of agricultural depression. It was against the fluctuations in supply and in price resulting from climatic hazards that the policy of stabilisation by import and export controls was directed.

The abolition of controls on food prices, of which the repeal of the Corn Laws was the final act, was not responsible for the long depression in farm prices. In fact the repeal of the Corn Laws was followed by a decade – 1853–62 – that was described as the golden age of British agriculture. It was only when the effects of the long distance transport systems built by the British Industrial Revolution were realised in the extensive opening up of the virgin lands of the New World that agriculture went into the great depression. Thus the real cause of the depression was, as R. A. Fisher pointed out as early as 1929, the production of food in excess of the ability of urban populations to consume it. Turning now to modern times, the good agricultural prices of the 25 years since the war have been due to a shift in the narrow balance between food supply and food demand as a result of population growth. The price support system devised during the war, and continued in the years following, has been the means whereby this changed balance between supply and demand has been reflected in farm prices. It is not the prime cause of good prices.

The way in which the support system has been operated has had a substantial effect on the pattern of agricultural development, and this must now be described. The origin of the support system was in subsidies to keep down the price of food to the consumer during the war. This was part of the wartime food policy planned under the influence of Boyd Orr's (1937) report on the relation between income per head in a family and the nutritional adequacy of the family diet. The importance of this report cannot be overemphasised. Here was set out explicitly the basic problem of food supply in modern urban communities. Even when food is cheap, the poorest sector of the community has not the money to buy a diet adequate in quantity and balanced in nutritional quality. Indeed poverty is a more intractable problem than agricultural productivity in feeding the peoples of the world.

The Ministry of Food, under the scientific advice of Drummond, seized the opportunity of wartime controls to plan a fair allocation of the food supplies available, by rationing and by establishing a balance between wage rates and food prices that would put the rations within the reach of all. Food at prices the people can afford became part of the conception of the Welfare State in the post-war period. The system of subsidies for British agricultural produce was carried on as a necessary part of the post-war development during the period when imports were difficult to obtain or pay for. It was later continued as the established procedure for the maintenance of a sound and vigorous agricultural industry as the needs of the community and the circumstances of agriculture changed. It is a feature of all human activity that an established system only changes slowly, even though the needs it was set up to meet change rapidly. Thus, under the support system, there has been a continuing tendency to maintain the incentive to produce those classes of food that were urgently needed in wartime circumstances. In particular, support for cereals has throughout been more generous than support for meat. This has continued even though cereals have become freely available on the world market, while with growing affluence, the increasing demand for meat has only been met with difficulty.

In assessing the impact of support on farming practice it is necessary to devise some statistical estimate of the degrees of support accorded to different agricultural products. Data are available in successive White Papers on the *Annual Review and Determination of Guarantees* (HMSO). A suitable statistic is the support accorded to a commodity as a percentage of the unsupported value of that commodity. The allocation of sums paid under price guarantees is straightforward. Partition of grants and subsidies can only be made on a rough allocation, but considerable changes in the allocation would be necessary to alter appreciably the overall figure, so a rough allocation is adequate (the actual partition used is set out in Table A in the Appendix). The unsupported value of the commodity is obtained by subtracting the total price guarantee payment on that account from total farm sales. The magnitude of support for the commodity is the amount of the guarantee payment plus the allocation from grants and subsidies to that commodity. Estimates of support to broad groups of commodities are set out for the decade 1959–60 to 1968–9 in Table 7. Farm sales are given for cereals separately from other farm crops for the years 1962–3 and subsequently. They are accordingly given separately in Table 7 when available.

The limitations of the Table must be recognised clearly. The system of guarantees and subsidies is not the whole of the

TABLE 7. *Estimates of levels of support accorded to agriculture as a percentage of unsupported values 1959-60 to 1968-9*

Year	Milk and milk products	Fatstock and wool	Eggs and poultry	Cereals	Farm crops		Misc. and valuation changes	Total
					Other farm crops	All farm crops		
1959–60	9.1	24.1	17.3	–	–	43.2	7.9	20.6
1960–61	10.0	23.4	11.4	–	–	50.4	9.5	20.8
1961–2	9.9	41.8	8.8	–	–	55.6	13.1	27.1
1962–3	6.1	36.0	11.1	(67.0	6.6)	37.0	17.3	22.8
1963–4	5.1	27.2	10.4	(95.5	7.1)	49.0	15.0	21.2
1964–5	4.5	19.8	16.4	(58.3	6.3)	34.4	13.0	17.4
1965–6	3.9	19.5	8.6	(36.0	10.4)	25.2	20.4	14.8
1966–7	3.5	17.8	8.4	(40.0	6.3)	24.1	24.8	14.1
1967–8	3.9	23.7	8.9	(30.6	5.5)	20.2	24.0	15.4
1968–9	3.8	19.7	7.2	(46.7	9.9)	30.1	26.8	15.5

support for agriculture. Liquid milk, being a perishable product, enjoys a natural protection against imports that can be deliberately maintained if necessary, and a price can be fixed by balancing consumer and producer interests, without considering import competition. The price of sugar is controlled under agreements with Commonwealth sugar producers as well as British producers, and the financial consequences of the agreements do not appear in the price review papers. Support for horticulture is a different basis, and figures for horticulture have been excluded from Table 7. Similar limitations apply to estimates of support or protection for industry and trade also, and it is only possible to set out a broad picture for comparative purposes. On this basis the figures in Table 7 are instructive.

The overall level of support ran at just over 20 per cent from 1959–60 to 1963–4, with a peak at 27 per cent in 1961–2. It then dropped and has levelled off at about 15 per cent. The fall has been due to some rise in import price levels for both meat and cereals, to a reduction in guaranteed prices for cereals, and to some tightening up of the price review agreements, for example the establishment of standard quantities to avoid an open-ended Exchequer commitment.

Between commodities, however, the degree of support varies very greatly. Milk, and eggs and poultry enjoy low, and on the whole declining levels of support. This is partly, but not altogether, due to the natural protection they enjoy, together with a quarantine protection. But it is also partly due to their progressive and economically advanced production practices – the Milk Marketing Board low cost milk scheme, the artificial insemination service, and the Board's good marketing arrangements; the efficient broiler industry, and the advanced technology of egg production. Fatstock have enjoyed greater support. In fatstock marketing uncertainty has in the past been great and the stabilising effect of the guarantees is reflected in the fluctuations from year to year in the rate of support.

Cereals have consistently enjoyed a very high rate of support, and it is important to bear this in mind in considering production policy. Cereal farming is the most profitable of English farm enterprises, even though the world cereal supply is abundant. Fatstock is relatively unprofitable, and a larger supply would be welcome. Yet the support level has always been higher – and usually very much higher – for farm crops than for fatstock. The greater part of the support for the arable sector of British farming is given as support for cereals. Potatoes and sugar-beet are cared for in other ways, but their management includes a substantial element of acreage control. Hence, in addition to the tendency to extend cereals at the expense of livestock resulting from the more generous support for cereals noted above, cereals have been extended disproportionately within the tillage sector. It is not surprising that the search for profitable crops to break a long cereal sequence has been unrewarding when cereals are supported and other crops are either subject to acreage restrictions, or are not within the support system.

Thus the operation of the British agricultural support system can be criticised in detail, but it is clear that adjustments could be made in it to meet changes in policy objectives. For example, if it became a matter of policy to encourage a return to mixed farming in order to solve the effluent disposal problem on intensive livestock enterprises, and to add organic manure to some of the specialist cereal acreage, appropriate changes in the balance of support between cereals and fatstock could readily be made. In this sense the support system is a more flexible instrument of agricultural policy than a system of tariffs would be.

In terms of level of support, the rate is modest. The rate as a percentage of unsupported value may be regarded as roughly the equivalent of a tariff at the same rate. The overall figure for agricultural support running from 22 to 14 per cent, with one high figure of 27 per cent, is similar to the range of tariff support given to industry (see for example, Anon. (1966) *The Common Market and the United Kingdom*, which gives an average import duty for all industrial products in 1962 of 18.4 per cent). The difficulties of the farming industry in recent years suggest that a support rate of 14 or 15 per cent is too low for satisfactory operation, but a rate of support comparable with the rate of protection afforded to industry might be maintained as a standard. Thus support for agriculture can be seen as no more than equal treatment with industry, while the form in which it has been administered has made it a tool of policy consistently more effective than tariffs, and having the potential of an extremely sensitive policy instrument.

Consideration of the British agricultural system would be incomplete without reference to the agricultural systems that were generated by the impact of the British Industrial Revolution on the New World and on Oceania. They began as exploitive systems, making agricultural use of virgin lands to grow agricultural produce for export to Britain. They suffered, as British agriculture suffered, from the long depression. Moreover, the communities in which they grew up followed the path trodden earlier by Britain in industrialisation. Thus the United States of America changed over a period of 160 years from a rural community with 94 per cent of the population engaged in agriculture to an industrial nation in which only six per cent of the people are on the farms.

The United States has gone further than any other of the new agricultural countries in industrialisation and in the reduction of the numbers engaged in agriculture. Canada and Australia have gone a long way, New Zealand not so far. The development of industry has gone on under the shelter of protective tariffs, and these have had the effect they were intended to have, of making it difficult for British industrialists to maintain their position in the markets of the new countries.

The process of development has been essentially one of building up balanced economies in the new countries by a redeployment of the population out of agriculture into industrial occupations. Since modern agricultural technology is such that it is possible to produce all the food and raw materials supplied by agriculture and forestry with less than 10 per cent of the population employed on the land, this redeployment is a necessary part of the creation of higher standards of living. It is a common feature of human progress, however, that the initiation of new enterprises is easier than the abandonment of the old. Even the United States, which has spent large sums in compensation for retirement of land from cultivation, has not balanced her agricultural production with the needs of home consumption and overseas exchange, and is generally in a position of surplus over what can be profitably exported. In Australia and New Zealand, the spirit that motivated the first agricultural development of these countries still persists, and there remains the urge to open up new land, even though agricultural production is sufficient, and the threat of unsaleable surpluses is always at hand.

The western world has yet to learn how narrow is the margin between surfeit and hunger. For half a century western farmers suffered the depression because of an excess of production – not a very large excess, but one that weighed down an inelastic market. For a quarter of a century they have enjoyed prosperity because the rate of increase in demand was as much as they could match with increased production. The time has come now to seek consciously to establish a balance between supply and demand, with variations due to climatic uncertainty buffered by adequate, but not burdensome reserves. This is the more important because, in some areas and under some systems, western communities are in danger of doing permanent damage to the land from which their resources in food and industrial crops are derived. In the early decades of this century attention was drawn to the enormous erosion damage resulting from the application of British farming patterns to regimes of land and climate in the New World to which they were unsuited. Systems have now been devised by which erosion can be contained, and the reclamation of eroded land has to a large extent been mastered. New hazards have now appeared. The indiscriminate use of pesticides and herbicides has had unexpectedly serious effects on the ecological balance in which agricultural systems are practised. Effluents and by-products have caused pollution in sources of water required for other purposes. The development of great urban concentrations has given rise to a load of waste that almost overwhelms the biological systems in rivers and poses a vast disposal problem and generates an enormous litter nuisance.

Advanced western communities are pressing on their environment even more heavily than the multitudinous but poor populations of Asia. It can be argued that there is still room, that the problem is not numbers but distribution. This is not tenable. Any society needs for security an environmental margin of safety. Storage gives security against year to year climatic hazards. Only a margin of unexploited resources can give assurance against long term climatic change. Damage to the environment is inevitable as man experiments with new materials and new techniques. Recovery is possible, but only if there is a margin, and the damaged resources can be put out of production for regeneration. Western communities are like Allan's (1965) husbandmen practising shifting cultivation who have so encroached on the regenerative phase that a progressive decline in productivity becomes inevitable.

The heritage of the Victorian age is the urge to exploit every resource that becomes available. It was a defensible philosophy while there were still unexplored resources beyond the frontier. In the second half of the 20th century the limits of our resources can be seen, and where our fathers exploited we need now to conserve.

CHAPTER 8

SUB-SAHARAN AFRICA

Agricultural circumstances are difficult over most of Africa. The continent is an ancient land mass, and the soils are in general poor in nutrients and often also in physical state. The wide distribution of tsetse and associated trypanosome diseases imposes a major limitation on cattle keeping and a not inconsiderable limit on the distribution of human populations. Rain-fed agriculture is only beginning to emerge from the early exploitive phase characteristic of a primitive farming economy. Shifting cultivation is widely practised, and cattle rearing communities are largely nomadic. Settled agriculture has scarcely got beyond the stage of subsistence farming, except where a cash crop has been added to it, and livestock keeping is not integrated with crop production. Shifting cultivation, nomadic cattle herding, and settled farming constitute three distinct farming systems, which call for separate consideration.

Nomadic herding is the simplest system, and is conducted in the poorest and most difficult environment. Its basic essentials are that there shall be grazing and water for stock in some part of the territory occupied, at all times of the year. So there are regular seasonal migrations, as the fodder and water situations determine. Such a system is subject to strict geographical limitations. There must be widely different climatic regimes within walking distance of each other if cattle are to be provided with forage and water in all months of the year.

Geographical circumstances favouring nomadic herding are found in various parts of Africa. Two situations thus exploited in the Sudan are those of the Red Sea coastal area and of the Sudd on the White Nile. In the Red Sea area, the Haddendowa tribe practice transhumance between the coastal region and the hills of northern Ethiopia according to the season. In the Sudd region of the White Nile, water and grazing are available on high ground during the rains and on the swamp margins as the White Nile flood recedes in the dry season.

There is no sharp demarcation between nomadic herding and systems involving crop production. In the Sudan the Dinka, for instance, depend substantially on crop production for their subsistence, and they are settling into a more crop dependent farming system. Nevertheless, they keep their cattle, and the herds and herdsmen practise transhumance as they used to do. In Kenya the Kipsigis, fairly recent cattle owning immigrants in favourable territory, are rapidly developing a mixed farming system copied from European farmers, by the enclosure of tribal lands. The Kipsigis have gone faster than the Dinka, having the example of European farms on which many of them worked, and being more favourably situated in relation to markets, but the trend is the same, and is the consequence of cattle owning tribes occupying land capable of further development.

Improvement of nomadic herding has come first from protection of cattle from epidemic disease. The immediate result has been overstocking, as change in social and ceremonial values lags behind changes due to improved technology. Improvement of water supplies by the provision of wells and tank dams is the next step in development, as it makes possible a better distribution of the grazing load. Increasing cattle numbers on better managed grazing will only provide a better living if the offtake is increased to make possible the satisfaction of new needs. Thus incentives to sell must be stimulated and new opportunities such as education and consumer goods provided. Ultimately, the objective in pastoral areas must be a managed ranching system.

The second major agricultural system is shifting cultivation. It is characteristic of areas with a reasonably good rainfall but a very poor soil and no input of fertility from outside the system. In this respect it is worth noting the similarity between the poor soils of the coastal plain pinelands of the Carolinas and Georgia in the United States, where a stable agriculture was only really established when fertilisers became available, and the leached sands of the miombo forests of the Congo and Zambia in Africa, where shifting cultivation is still practised.

The practice differs in detail in different parts of Africa. In Zambia, the Chitimene system studied by Allan (1965) involves the concentration of the elements of fertility. Vegetation is cut from a large area, piled on the small area to be cultivated, and burnt. In the Congo, the very long rest necessary for the recovery of poor soils under forest was systematised by the Belgians, who established a rotational clearing and regeneration system involving as much as 30 years regeneration after two years of cultivation. On the grass savannahs of Kordofan in the Sudan, the practice of 'harig' involves burning *in situ* the residues that have accumulated after some years of grass growth. There, the cultivator's chief problem is heavy weed growth and not low fertility status. By postponing the burn until after the first rains, the weeds germinating with the onset of the rains are destroyed. The delay means late sowing, and results in reduction in yield per acre, but yield per unit of labour is increased and since land is not limited, return to labour is the important consideration.

Shifting cultivation can only be transformed into settled agriculture if the restrictions on continuous cultivation are removed. Where these are primarily weed growth, as in Kordofan, mechanised agriculture could be introduced if returns justified it, and indeed mechanised rice growing has been attempted with some success on river flood plains in the south west of the Sudan. Where, as in most shifting cultivation areas, nutrient poverty is the limiting factor, the only answer to shifting cultivation in the long term is fertiliser use. Meanwhile, shifting cultivation will continue and the prospect of a higher standard of living will depend on the earnings of labour outside agriculture. These areas have been major recruiting grounds for mine labour in southern Africa and the export of labour, particularly male labour, is likely to remain their only important source of money income. This gives rise to problems, both social and agricultural, from the prolonged absence of a large part of the male population.

Shifting cultivation is only practised out of necessity. Clearing forest for short term cropping is heavy labour, and land once cleared is farmed for as long as it will produce economically. So there is a gradation between poor land where a long term rest is imperative, better land on which cultivation can be maintained for some years and only short rests are necessary, and the best land that can be kept permanently under crop. Settled agriculture developed wherever there was the combination of good soil and reliable rainfall. Originally it was subsistence farming based on the African cereals – sorghum, finger millet and bulrush millet – in areas with a long dry season, and on starchy bananas in East Africa and yams in West Africa in areas with a double rainy season and short and mild dry seasons. Maize, cassava and sweet potatoes and other New World crops were added when they became available. Pulses were also grown, the indigenous cowpeas, *Dolichos* and Bambara groundnut, and more recently the New World groundnut and *Phaseolus* beans.

Development in Africa was only possible if African countries could produce something that would bring western money into them. Nomadic herding offered no prospects in this direction. Shifting cultivation offered only the possibility of drawing off male labour for other activities, for instance mining. Settled agriculture provided the possibility of adding cash crops.

The concept of the 'cash crop' is peculiar to areas that were until recently restricted to a subsistence economy, and where there were the resources of unused land and under used labour that could be devoted to new kinds of agricultural production. The incentive for production was the opportunity to acquire those extra commodities and services that the western world offered, and which were only obtainable in return for western world money. While the food crop farming system remained unchanged, a new range of crops was added to the farms, the produce of which was sold to outsiders for the outsiders' money. The money was used for the purchase of services such as administration, security, education and communications, and for goods such as cloth, bicycles, galvanised iron, sewing machines and so on. The more important crops so introduced were cotton, coffee, tobacco, cacao, oil palm, cola and sugar-cane.

It is important to note that the basic step in the modernisation of African agriculture was the straight imposition of a cash crop on a subsistence economy with settled farms. This involved more land and more labour but not a new system. Land was at that time adequate. The extra labour came in general from the diversion of male activities from defence to agriculture. Hence, the common complaint one used to hear of the idleness of the African man. It was his function to be idle, to be ready for a fight, either offensive or defensive, or to put in a limited spell of clearing when his wife wanted new land. It was never specifically stated and accepted that cash crop production was in fact the modern equivalent of his old occupation, since it paid the taxes that financed the administration that secured him against tribal war.

The importance of the cash crop in the development of African economies cannot be overestimated. Cotton financed the railway, lake steamers and roads in Uganda. Cacao financed similar developments in Ghana. It was natural therefore that the developing agricultural departments should pay disproportionate attention to the crops on which the stability of the new economy depended. Food crops were allowed to take care of themselves except for insistence by the administration on adequate reserves against famine. The indigenous food crops had the advantage of a long established balance with the local environment and introduced ones such as maize, cassava and sweet potatoes, either established themselves, or disappeared according as to whether they fitted into the local agricultural pattern. Cassava mosaic virus was probably as serious in its effects on cassava yields in Uganda as cotton leaf curl virus was on the yield of cotton in the Sudan, but the attention it attracted from research workers was infinitesimal by comparison. It should be remembered also that the development of cash crops under Colonial regimes in Africa went on during the long agricultural depression. Food was overproduced in the world, and the Colonial agricultural departments rightly limited their interest in food crops to ensuring adequate production for internal consumption.

In some African territories, in addition to the introduction of cash crops to stimulate the development of a cash economy, there was an attempt to transplant European type farming into the African environment. The success of European farms and plantations depended on the suitability of the climate for the purpose. This was quite satisfactory in the Kenya highlands, highland areas of the Congo, and the warm temperate climate of the more southerly territories, but not in West Africa. Its impact on African farming was twofold. First, the planters introduced and established exotic crops some of which – e.g. *arabica* coffee, cotton and sisal – were readily taken up by African farmers, and some which – e.g. rubber, tea and sugar – were adopted more slowly, and with attendant problems of capitalisation and processing. Second, European farmers, particularly in Kenya, taught Africans, chiefly by example on their own farms, the merits of European farming systems.

The degree of modernisation achieved under Colonial rule was no more than a part way stage. Roughly, it could be said that subsistence was little changed, except by the introduction of New World crops, such as maize, cassava, and beans. The growing need for cash to raise the standard of living in the social and material aspects of life was met by the export of crops from the better farming areas and man power (for the mines) from the poorer. This intermediate form of agriculture is no longer adequate. On the one hand, in some areas populations have increased to the point where there is no longer land to spare, and indeed pressure of man and stock on the land has led to a serious decline in productivity. On the other, the demand for a steadily increasing standard of living cannot be met by this agricultural hybrid. The next step is conservation, as a preliminary to intensification. This takes many forms.

Two examples may be given. In Kenya, the African Land Development Board (Anon. 1956) undertook very successful conservation measures in areas of uncertain rainfall and showed that existing human populations and adequate stock populations could be maintained on land that had been heavily overworked, if proper controls of grazing and erosion were established, and holdings were rationalised. In parts of Kenya, there is a spontaneous move towards individual holdings with individual rights in land, and the advantages in land management and soil fertility of individual control can be seen on many African farms. In this respect, the political accident that established individual freehold in large areas of Buganda (Uganda) have turned out greatly to that country's advantage. In West Africa, the agriculture of the cacao crop has been exploitive, the forest being progressively destroyed and the cacao in turn falling a victim to pests and diseases. Conservation has come first by way of insecticides and fungicides, protecting the trees and improving their yield performance. On information now available through research, yields can be raised by better agronomic practice, and with this and with pest and disease control, fertiliser practice becomes worthwhile and a new standard of productivity is within reach of the cultivator who is prepared to organise his own work to new standards.

As the developing countries build up a balanced economy with large urban populations, there arises a market for food crops. The distinction between food crops and cash crops disappears, and all sectors of the farming system contribute to the farmer's cash income. This is occurring most quickly and most obviously in the vicinity of growing towns and mining areas. Large populations earning their living from activities other than agriculture have been built up, and they have become purchasers of food for money. The mining areas of southern Africa initiated this development, and in those areas there arose a place for European farmers to grow food for mine labour. Further north, the urban developments of East Africa, though fed partly by European agriculture in the Kenya highlands, draw most of their supplies from the African farmers in the immediate vicinity. Indeed, in Uganda the ability of the local agriculture to meet a growing local demand for food was so good that the development of the mechanised Busoga Farms to feed Jinja when the Owen Falls dam was built was a costly failure.

The entry of food crops into the cash economy does not always come by way of urbanisation. Successful development of a cash crop may lead to a specialised form of agriculture in which some farmers grow the cash crop and some grow food to sell to the cash crop farmers. This happens in most plantation systems of agriculture, as for example on farms around sugar estates, but it has also developed in West Africa where the savannahs of the Middle Belt feed the forest zone to the south, in which the great cacao exports are grown.

In most of Africa the work of providing enough food to eat is not very onerous, and few people suffer hunger, except possibly for a month or two before harvest, or where much of the male population is away in mines. No great agricultural revolution would be required to supply the extra to fill that gap. The major African problem is malnutrition – a poor balance in the diet. This is more a matter of ignorance of human requirements than of inability to meet them, and with education the demand for protein foods in particular is likely to increase greatly. The best means of meeting this demand is by integrating livestock into the farming system, improving local stock by better feeding and better breeding, and using stock as one tool in the campaign for the improvement of fertility levels. This integration is likely to lead to a farming system more like the mixed farming of Britain. For stock, fencing is required, and forage supplies must be improved and controlled. There is then also the opportunity to develop a manuring system. In fact there are in mixed farming the essential components of a conserving and improving system of farming.

Further agricultural progress in Africa will come from developing the food crop sector of farming to meet the increasing needs of local towns. Cash crops are well established, and their exploitation as far as the limited market demand permits, is obviously necessary. The greater part of the research work that has been done was concerned with their improvement, and there is a great store of knowledge of them that has yet to be applied. The position is in many ways similar to that in British agriculture in 1939. The western world was over supplied with agricultural produce and there was no incentive until the last war to apply the knowledge then available, since it would lead to an increase in what was already in surplus. Similarly, the world now has a sufficiency of all the tropical cash crops and a surplus of some. Some producers, for example those of rubber and coffee and cacao, already have experience of the fall in return to the producer that may follow a rapid rise in crop production. So the application of knowledge to the production of cash crops can only be profitably undertaken at a rate that will keep production in reasonable balance with demand.

In the improvement of production in the food crop sector of African agriculture, the basic problem is the poverty of the soil. Shifting cultivation is widespread, and in many areas of settled farming there is in fact recognition of a need to allow some rest of the land from time to time. Both the regular shift and the occasional rest are imposed by the low level of fertility of the soil. Where the rewards of agriculture are greatest, in the neighbourhood of the towns, intensification is attempted, but its success is limited, and the evidence of the decline in productivity that follows is commonly to be seen in poor farming and poverty stricken secondary vegetation.

This low fertility status is a common feature of the soils of sparsely populated developing countries. All the nutrients needed can be supplied where there is the technology and the industrial plant to manufacture fertilisers. More research is required on needs and responses in particular circumstances of soil and climate, but in the long term, in areas of good rainfall the productivity of tropical lands is a matter of nutrient status. With the limited exception of volcanic soils, with reserves of minerals to be released by weathering, and alluvial soils, to which there are accretions of fertility from the products of erosion, nutrient status will only be raised by fertiliser application.

Since tropical countries cannot look forward to any great increase in the sale of cash crops to enable them to buy industrial goods, they cannot look to the fertiliser industries of Europe and America to make good the nutrient deficiencies of tropical soils. They must ultimately have their own fertiliser industries. Raw materials are available, at least in some of the African countries. In Uganda, for instance, there are phosphate deposits. Sulphur, which is locally important, is being discharged into the air from the copper smelter. Atmospheric nitrogen is free, and power is to be had from the Nile.

Three requirements remain: capital equipment, technology and an adequate market. The short term solution is to avoid all three by importing such small quantities of fertiliser as will meet the current demand. This is a beginning, but if it were successful and the demand were to build up, the foreign exchange position would become very difficult. This can be seen in India where although there has been heavy investment in the fertiliser industry, fertiliser imports have risen faster than home production, creating a heavy burden on limited resources of foreign exchange. At some stage a local industry must be established as in India. If capital is to be provided out of foreign exchange earnings, foreign exchange must be carefully husbanded. If there is to be any hope of manning the industry without heavy and expensive dependence on expatriate technologists, technological education must be developed in terms both of numbers trained and of quality of training. There remains the problem of the market. Western enterprise, in a context of expensive labour and cheap capital, has exploited economies of scale to the extent that nothing but a vast fertiliser plant can be regarded as economically competitive. So nowhere is it economic to build a fertiliser plant, unless there is a vast market for its products. In the absence of a large supply, no large market has been built up.

This is a situation with which the developing countries have long been familiar. Many instances could be cited of delay in the installation of such development essentials as ports and railways, because there was not the traffic to make them remunerative and the traffic could only be developed after the facility had been provided. In the same way there will never be a large market for fertilisers until there is a local fertiliser industry. In view of the low nutrient status of tropical soils, strategy dictates that there shall be a fertiliser industry. It is a matter of tactics to strike a balance between the higher production costs of a small industry and the longer period of under production that would be incurred by a large one.

Fertiliser production has been considered because of the importance of nutrient status for the exploitation of tropical lands. It is also a good illustration of the problems involved in organising a community that will have at least a degree of independence of the great industrial states. At present in most African countries it is not possible to have a motor car, a sewing machine, or even a hoe or a panga without selling cacao, palm oil, cotton, or some other tropical crop, in a temperate region market. This situation will persist for a long time and it is imperative that since the market for these crops is in a continuing state of near saturation, the African countries should so plan their economies as to conserve their receipts for export crops and use them for those imports that are vital for economic growth. Wheaten bread, along with temperate region butter and cheese and tinned meat, can be bought in all the towns and many of the larger villages throughout Africa, and improved nutrition comes to be associated with dried skim milk provided through overseas aid. Even poultry production and dairying to meet the needs of the cities of the West African coast rely on imported concentrate feeds. All this involves misuse of scarce foreign exchange, and starves local agriculture of support for just those sectors of farming that could lead to the sustained growth of modern systems, managed by progressive farmers. The real opportunity for the literate farmer is the town market. It should be national policy in all African countries to reserve it for him. That he can exploit it is shown by the success with which men with education and with experience in town occupations have gone into food farming to serve the Kampala market in Uganda. That he should have it instead of the importer is dictated by the national need to reserve foreign exchange for other things.

The potential of the town food market, and the needs of town populations, should be appreciated. Urbanisation the world over leads, at least initially, to a fall in nutritional standards. Families which have grown their own food and have had access to the wild where greens for spinach may be gathered and fruits and berries picked in season, find themselves in the position of having to buy all their food. Not only can they not afford more than a basic fill-belly diet, but they are not aware of the value in terms of vitamins and minerals of the minor accessories and garnishes that they have lost. Worse still is the plight of the single person migrating to the town in search of a cash income and often under an obligation to save and make remittances home. These immigrants to the towns form the core of the urban malnutrition problem, a compound of poverty and ignorance, and little is being done to help them.

The exploitation of the town food market calls for agricultural development in three sectors, the production of staple foods, the provision of horticultural produce, and the development of a dairy and poultry industry. Staple food production is more than a direct intensification of the food production patterns of subsistence agriculture. The needs of the towns call for emphasis on storable types of food and for those that serve the changing food habits of town dwellers, such as restaurant eating and the preparation of meals to be taken out to factory or office. Vegetable production is of great importance in meeting the dietary losses that occur when farmers become town dwellers and no longer have available the accessory foodstuffs that they easily get in the country. In most African countries horticulture has not had adequate attention from agricultural departments. Even check lists of African vegetables are few and hard to come by and research on vegetable production has been very limited.

Dairying and poultry keeping have fared a little better. Veterinary departments have in many countries established at least one dairy enterprise near the capital city and they have distributed cockerels and more recently day-old chicks of productive poultry breeds. Successful dairying in the tropics depends first on preventive veterinary medicine, second on good feeding and third on breed improvement. Once security from disease has been achieved, regular and adequate feeding must take the place of the customary regime of surfeit in the rains and semi-starvation in the dry season. The African cow does not give a return for more than a very modest level of feeding, and when the food supply is assured, breed improvement must begin. There are productive tropical breeds, such as the Sahiwal and Red Sindhi in Pakistan, the Kenana and Butana in the Sudan and the Criollo Lechero in Latin America. The difficulty with all of them is that the number of animals of known history and with reasonably reliable production records is so small that they cannot now be used as a basis for a developing dairy industry outside their own particular areas. So the breeding of productive dairy stock must depend on the introduction of high genetic potential by crossbreeding with temperate region breeds. In some areas indeed, the direct introduction of European breeds is possible. The establishment of a high standard of veterinary preventive medicine in Kenya and Uganda has led to the establishment of a productive and growing dairy industry using European stock. This does not mean large and continuing livestock exports from Britain. The objective in most areas must be the establishment of new breeds, like the Jamaica Hope, combining the production characteristics of temperate breeds and the hardiness, disease resistance, and feed economy of tropical stock.

Poultry production is probably easier to organise than milk production. Eggs do not pose the same public health problems as milk, and modest improvements in feed and housing make possible the introduction of superior breeds. Moreover, day-old chicks can be sent easily and without hardship or loss from Britain to the ends of the earth by air. This is all to the good, though one must hope, in the interests of conservation of foreign exchange, that the import of stock that could be hatched at home will be a passing phase. What is less desirable is a tendency to initiate factory farming where its main justification, the high cost of labour, does not hold. Free-range poultry may be subject to risk from disease and parasites but in a hot country with comparatively cheap labour, simple housing and a modification of a straw yard or deep litter system is as far as intensification need go. To import feed by sea from Britain, as is now done for some enterprises in West Africa, is beyond all reason in terms either of national economics or of nutritional standards.

The production of food for the growing towns and particularly quality foods – vegetables and dairy and poultry produce – is the new growing point for the agriculture of African countries where land is still to be had by the ambitious and enterprising. Here is the opportunity for the school leavers who want something better than subsistence and an export crop. Here is the opportunity to strike at the root of malnutrition. Feeding the towns is not a gift but a challenge. It will take imaginative research to develop from local resources a diet fitted to the needs of townsmen, and wise and firm administrative action to divert the housewife from imported goods to local produce. Moreover, the plain fact is that good diets cost more than poor ones and the spread of good nutritional standards to the poor among townsfolk will depend on rising wage levels as well as on nutrition education and the availability of supplies.

Africa is a continent of abundant land, much of it enjoying adequate and reasonably reliable rainfall. Population growth has hitherto been limited by the intrinsic poverty of the soil, in some areas by the absence of dry weather water supplies, and by disease and political insecurity. Control of disease and the establishment of political stability have resulted in population growth, and a demand for more of the fruits of the soil, both for subsistence and for sale as a means of raising living standards. Agricultural research has led to the identification of soil nutrient deficiencies and the specification of fertiliser mixes and farming systems whereby soil fertility can be raised. The stimulus of substantial rewards for entry into the commercial sector of farming has led to the beginnings of modernisation and the establishment of the exchange between town and countryside on which farming progress depends.

Thus in Uganda, money was introduced to the farming community early in this century through the cotton crop. After the second world war, the development of the *robusta* coffee crop provided opportunities that were seized by profit-minded individuals – some from the land-owning and farming sector, and some from business and government employment – to establish large farms and to amass considerable wealth. More recently, the growth of industrial and service activities in Uganda's towns led to the expansion of food crop farming for sale to the towns, and to modern enterprises in dairying and poultry production. As this development has gone on, inputs of fertilisers, pest and weed control chemicals, and power and machinery, have increased as knowledge became available from research stations and supplies were acquired either from overseas or from newly developed local resources. Indeed, the rate of development of agriculture has in most sectors been equal to the rate of development of effective demand. There is still room for an increase in dairy production, since the rate of increase in productive stock is slow, and determines the rate of expansion of the industry. But in vegetable production, in the supply of the basic indigenous foodstuffs, and in poultry production, supply keeps pace with demand.

Here is the challenge to planning, to hold the narrow balance between hunger and surfeit, and to develop agricultural potential as population expansion and nutritional needs dictate.

CHAPTER 9

INDIA

Agriculture has been practised in the Indian sub-continent for 5000 years or more. During at least the last 2000 years, the population has been large enough to press heavily on the better farm lands. The current agricultural situation is dominated by the effects of long continued cultivation with only haphazard return of nutrients to the soil, and a comparatively simple system of husbandry. Yields are low, but a base level of fertility has been reached and there is no evidence of any further secular decline in productivity. There are exceptions to this general situation where under irrigation the water table has risen and land has been lost by salinity, and where on the banks of rivers and on steep slopes gully erosion is active.

India is subject to the risk of famine through crop failure in years of poor monsoon rains. Nevertheless the agricultural system is stable and productive, and the reliability of rainfall is as good as in other countries where stable farming systems have been established. The risk arises from the narrow margin between productive capacity and the needs of the vast Indian population. Where there is heavy pressure on biological resources, a reduction in rainfall within the normal limits of expectation can create conditions of shortage, or even of famine.

The climate is warm temperate to tropical, warm enough for crop growth throughout the year. Even in northern India where there are occasional frosts in winter, it is only at high altitudes that the cropping season is limited by low winter temperatures. Cropping patterns are dominated by water supply, but the effect of temperature is still considerable. For instance, wheat (*Triticum*) and Bengal gram (*Cicer*) are *rabi* (cool weather) crops, whereas jowar (*Sorghum*) and red gram (*Cajanus*) are predominantly *kharif* (summer monsoon) crops. Whether single cropping or double cropping is practised is determined primarily by water supply. Where irrigation water is available double cropping is common. On rainfed lands, there is rarely water for two crops on the same land. Nevertheless, crops may be grown in both seasons, the *rabi* crops being dependent on water stored in a retentive soil during a monsoon fallow.

Rather less than 19 per cent of India's crop land is irrigated and calculations of available water and land have led to the conclusion that about 20 per cent is the maximum for which irrigation can be provided. This is the land that is most productive now, and that has the highest potential for improvement. In many areas, but not all, irrigated lands are less affected than rainfed lands by the vagaries of the monsoon. Irrigation water from monsoon fed rivers will fluctuate with the monsoon, and the reservoir storage available only offers a limited degree of security. In contrast, the waters of the northern rivers, fed by snow-melt from the Himalayas, are influenced by climatic factors other than those determining the monsoon, and irrigation in northern areas has consequently an extra safeguard. Security is greatest where water is obtainable from an underground source. The great northern Indian alluvium, for example, is an enormous aquifer, and the development of open-well and tube-well irrigation to supplement the supplies from canals has meant tapping a source of water that is substantially buffered against year to year fluctuations in the monsoon rains.

The great rainfed tracts of India are in the peninsula, south of the vast alluvium that lies at the foot of the Himalayas. The watershed is on the Western Ghats, not far from the west coast. The western slopes are steep and have a heavy rainfall. The eastern slopes are gentle and have a moderate to low rainfall. Irrigation, mostly for rice, is practised in river valleys and deltas to the limits imposed by the availability of water. The rest of the land is devoted to rainfed crops, predominantly in the *kharif* season. Where there are retentive soils, as for example the black cracking clays of Malwa and the Deccan, *rabi* cropping is practised on stored water. These rainfed lands of the peninsula are the areas of greatest risk. Population pressure is heavy, and the peasant economy is without adequate resources to build up reserves against a dry season.

In terms of the range of crops and stock that is embraced within the farming systems, the Indian sub-continent is one of the most remarkable areas in the world. Farming began in the region with the introduction of wheat and barley and associated crops from the Middle East centre of origin of agriculture, probably about 3000 B.C. By 2000 B.C., the indigenous crop plants of India, rice, pulses and cotton, had been domesticated and at about the same time there was an influx of African cereals. At an early date also India gained bananas and sugar-cane from South East Asia. More recently, with the arrival of the Portuguese, the introduction of American crops began, leading to the establishment of maize, American cotton, groundnuts, and potatoes. In livestock likewise the range of species is wide, but a larger proportion is of indigenous origin. The cattle and buffaloes used in India are native. Sheep and goats and camels were probably domesticated to the north west of the sub-continent. Pigs may well be indigenous, and the domestic fowl is a local domesticate. Thus India enjoys as great a wealth of domesticated species as any country, and the way in which crops in particular have been fitted together into geographical and agronomic niches to which they are suited makes the agricultural biology of India a study of absorbing interest.

The farming situation in India is very different from that in Africa, though crop yields have for long been of the same order. The low yields of African agriculture are the yields obtainable on intrinsically poor soils. Even these yields are not sustained without periodic – often very frequent – resting. The low yields of Indian farming are the result of 4000 years of exploitive agriculture. Few soils in sub-Saharan Africa would stand 20 years of the cropping intensity that has been imposed on Indian soils for 40 centuries. It is an index of the robustness and intrinsic fertility of Indian soils that they have maintained an equilibrium in crop production. Yields are low, but for many years they have not become appreciably poorer.

It is this that has made possible the establishment of a civilisation in India in ancient times, and its successful continuation to the present day. The cities and states of Indian history were built on the enduring fertility of Indian soils. The civil engineering – canals, roads and railways – the administration and the educational system of the British Raj and of Independence, likewise depend on this enduring fertility for their success. Pressure of population on the land is great and increasing. Up to the middle of the present century it was regarded as something that countries like India and Pakistan would always have to live with. Agricultural improvement went on slowly and the object of policy was to ensure as much food in the market place as there was money to buy. In areas of shortage this was achieved by transport from areas of plenty. There was little thought of making a radical change in food production, and no hope of eliminating poverty and providing adequate food for everyone.

Thoughts of an agricultural revolution followed the acceptance of revolutionary thinking about the whole social structure after the last war, with the realisation of the enormous disparities between countries in standards of human welfare. The people of India have come to believe that India ought to be able to get, and to get within a reasonable time, the benefits that had become freely available in the West. One of these benefits, that of preventive medicine, has become available, and has had the effect of reducing the death rate, greatly increasing the duration of life, and raising the rate of population growth. India and Pakistan ran into large food deficits and these were aggravated by bad monsoons. Serious shortages were only avoided by concessionary purchases and gifts of cereals from North America. For a decade it looked as though the Indian sub-continent would be overtaken by a Malthusian catastrophe.

Those who knew something of the agricultural resources of the region were aware of the unexploited potential that still existed. Indeed, the production of food grains in India was increasing steadily long before the bad monsoons of 1965–7 made the world aware of the gravity of India's difficulties. In the two bad years the total food grain production was still some 40 per cent above the level of production in the early 1950s. The 'infrastructure' for agricultural improvement in this period included an agricultural advisory service, a research service, and a considerable industrial sector devoted to agriculture. Improved crop varieties had been available from the research service for many years, and had had considerable effect, though the absence of a good seed supply system impeded progress. Fertilisers were available both locally and by import, and many sectors of the farm community were familiar with their use. With these resources there was an advance in Indian food grain production, erratic from year to year according to the vagaries of the monsoon, but leading to the doubling of production in two decades (1950–70).

Improvement at this rate was substantial, but it was not sufficient. The best that could be hoped was that food grain production would increase as fast as population increased, if there were no bad monsoons. Two bad years showed up the weakness of the situation, and fortunately at the same time there arose an opportunity to make much faster progress in some sectors of the agricultural economy. A new range of wheat varieties bred in Mexico, and of rice varieties bred in the Philippines became available, which respond to heavy fertiliser applications without lodging. Under trial in India it was shown that with the proper fertiliser application and appropriate agronomic treatment, they were capable of giving large increases over the yields obtainable from Indian varieties.

The decision to concentrate agricultural development work on high yielding varieties was not an easy one. Grown under normal Indian farming conditions they yield no more, and often less, than standard Indian varieties. They are indeed solely a means of exploiting high fertility conditions. To establish such conditions it is necessary to apply fertiliser generously, to improve agronomic practice, and to control more precisely than is customary the amount and timing of irrigation. It was argued, with justification, that a greater return could be obtained from the fertiliser available to India by applying it widely in smaller doses to the crop varieties in current use, than by putting it onto a limited area of the new high potential varieties, in heavy dressings.

The argument for the policy of high fertiliser dressings on high potential varieties was that it was necessary to demonstrate beyond any doubt, the feasibility of establishing, at least in a sector of Indian agriculture, a high productivity farming system. The policy was accepted, and wheat seed was imported from Mexico in bulk to start the programme in the wheat areas. The enterprise was successful and demonstrated the feasibility and profitability of heavy fertiliser use with good irrigation facilities and new varieties that can stand up under high fertility conditions. With wheat and with rice the 'package deal' has been shown to work, though in farming practice success with wheat has been greater than with rice. The nature of the package is important. It includes fertiliser, water, variety and the farmer. It is a mistake to regard this as a plant breeder's revolution. Thus the spectacular success in improving wheat yields in north west India compared with the limited advances achieved in the rice areas is due to the greater ease of adoption of better husbandry practices – fertiliser application, seed bed preparation, weed control and water management – with wheat as compared with rice. Varietal differences also entered into the matter. The new wheats proved to be more resistant to the Indian range of diseases and pests than the new rices, and the shortfall in quality was more serious in the rices than the wheats. But it was the husbandry factors that dominated the situation.

The improvement that has been brought about in India's wheat production of food grains justifies the term 'breakthrough'. The breakthrough generated a major change in a limited area. Its results were spectacular, but it raised a number of problems that will have to be solved if the full potential of the new situation is to be realised. Seed supplies have been provided through the National Seed Corporation. The entry of the new wheat and rice varieties to the market has been controlled by price support through government procurement.

Storage is a problem and is one on which wise policy decisions are particularly important. The first need for storage is to provide for the spread of consumption over the period between successive harvests. The second is to make possible a build-up of buffer stocks in good years to provide against bad harvests. India will become independent of foreign food aid when the country both provides enough home grown food in normal years and builds up a buffer stock that renders it independent of foreign surpluses in a bad season.

The breakthrough in wheat production raised the outturn of wheat by 50 per cent in three years. In the wheat growing areas of north western India, the agricultural circumstances were most favourable. The deep alluvial soil was responsive to applications of nitrogen and phosphate. There was an established canal irrigation system that provided a basic water supply, and there is a vast aquifer in the alluvium from which supplementary supplies can be drawn through tube-wells. This latter is of particular importance because it not only provides supplementary water, but it gives the opportunity to exercise the close control of water and nutrients necessary for high yields. The canal irrigation system was designed a century ago to provide some water at intervals to a large area as a security to low production farming. Under the new conditions, water is as critical as fertilisers for high production, and it must be applied with a timeliness and a precision that cannot be attained with the existing canal irrigation system. It was therefore most fortunate that there exists under the wheat lands of north west India an aquifer that can be exploited by individual farmers. The sinking of tube-wells has given the wheat farmer a degree of control of crop watering that could only have been achieved otherwise by a radical re-organisation of the canal irrigation system.

In north western India the farming community was pre-conditioned to react favourably to the new opportunity. Tradition had been weakened in two ways. On the one hand, industrialisation and urbanisation in that region provided more alternative opportunities for employment than exist in much of rural India, and on the other, there had been a great upheaval at the time of partition, with the loss of those who left for West Pakistan and the gain of those, many of them progressive farmers who had had substantial holdings, who came from West Pakistan to India. Even in the availability of land, circumstances favoured the wheat belt. With modern insecticides the malarial Terai was conquered and a substantial new and fertile tract was brought into cultivation, and the redistribution of the Indus waters made possible extensions to the irrigated lands of the Punjab, Hariana, and Rajasthan.

Conditions were not so favourable in the rice growing areas. Fertiliser responses were to be had, but water control in rice paddies is more difficult than in wheat fields, and cannot be organised so successfully by the individual farmer. Holdings are smaller and the pressure on the land has not been relieved to the same extent either by alternative employment or by an accession of new land. So the potential of the new rices has only been realised to a limited extent, and India's rice areas remain the areas of extreme poverty, and of social unrest.

Wheat in north west India and rice throughout the sub-continent are very largely grown under irrigation. If the break-through is to become a revolution, it must be extended to the great area of cultivated land for which no irrigation water is available. Production on these lands – 80 per cent of the total cropland – is determined by the same four elements, soil fertility, water supply, crop varieties and the farming community. On the rainfed lands, however, the farmer lacks the control that he has over irrigation water. Moreover, since nutrients are only available to a crop when there is soil moisture for growth and for root activity, fertiliser use is more uncertain on rainfed than on irrigated land. For these reasons, the impact of the ideas and practices of the 'green revolution' is limited over the great peninsula regions of India to those areas where irrigation is possible. One can see progressive farms, with high yielding fields of wheat and rice, wherever irrigation water can be provided in Malwa, Maharashtra, Andra Pradesh and Tamil Nadu. Indeed it is of great significance for the future prospects of Indian agriculture that these foci of high production, often small and dependent on wells or tanks, are to be found so widely. Nevertheless, all around them the vast areas of rainfed lands are so far very little affected.

A strategy for the improvement of productivity of India's rainfed lands must depend primarily on an understanding of the pattern and the variation of rainfall, and of the extent to which rainfall variation can be buffered by storage in the soil. New opportunities to influence the crop water supply relationship have arisen in two ways, in the amelioration of soil conditions and in the breeding of short term, fertiliser responsive varieties. Consider the consequences of long continued cultivation, with the limited power available from draft oxen, on some of the clay soils of peninsular India. The cultivated top soil, a few inches deep, has lost in clay content and become a sandy, impoverished and readily erodible surface layer, underlain by a heavy compacted clay layer with a very high bulk density. It is so dense and compacted that percolation into it is very slow indeed, and root penetration is greatly impeded. Hence run-off and erosion are accelerated, and as soon as the rains cease, drought supervenes. With the power now available to agriculture, it is possible to reconstruct soil conditions in such a way as to facilitate deep penetration of rainfall, thereby reducing run-off and erosion and increasing the soil water reserves available for crop growth.

At low fertility levels, production is greatest with crop

varieties that grow over the whole of the period during which soil water is available. Thus in areas such as Malwa, with retentive black cracking clays, the *kharif* (monsoon) crops are long term, growing on stored soil moisture for two months or so after the end of the rains. Fertiliser application is hazardous with such crops, both because the period is short over which nutrient/water relations are good, and because of variations in the date of termination of the monsoon rains, and hence in the amount of stored water on which to finish the crop. With the early maturing varieties now available from breeding stations, and with fertilisers, it should be possible to time the crop to grow and mature over the period of reliable rains, and to provide a fertility level over that period that will give high crop yields.

Swaminathan *et al.* (1970) have set out this strategy for the attack on the productivity of the rainfed lands. First is the definition of the period over which reliable soil moisture conditions can be maintained. Second is the breeding of crop varieties of the right duration to fit this period. Third is the application of fertilisers to establish nutrient levels adequate to ensure high crop yields over a short cropping season.

So far, only quantity of agricultural production has been considered. In terms of human nutrition, quality is almost as important as quantity. In terms of quantity, calorie needs rank first, and these are met primarily by cereals. When calorie needs have been met the next requirement is for protein. This is met in India chiefly by cereals, with supplementation by pulses, and by milk and milk products. Success in increasing cereal supplies has been offset by a decline in pulses. During the period from 1950 to 1968 when cereal production went up from 52 million tons to 84 million tons, pulses only increased from 8·5 million tons to 10·5 million tons, and fell from 16 per cent to 11 per cent of the total food grains. Over the same period progress in the improvement of dairy production was also limited. Thus the increase in food grain production has certainly not been accompanied by any improvement in protein/calorie balance. In fact it is probable that nutritional balance is poorer than it was before. The situation is mitigated somewhat in that the increase in production has been proportionately greater in wheat, which has a good protein content, than in rice which is poor in protein.

Plant breeding in India has been devoted disproportionately

to the cereals, but there is a considerable programme on the pulses. One means of improving the pulse situation is the breeding of very short term pulses that can be grown under irrigation between harvesting the *rabi* cereal and sowing the *kharif* crop. In addition, new varieties of *arhar* or *tur* (*Cajanus*) show promise of competing successfully with rainfed cereals, and may help to redress the balance. There are other achievements also in the improvement of pulse crops. Digestibility and amino acid composition can be improved by selection, and recently improved strains of *Lathyrus sativus* have been produced by selection which have a very low content of the substance believed to be the neurotoxin responsible for lathyrism.

Cattle are sacred in India, and it is commonly believed that this is a serious impediment to agricultural progress. Recently Raj (1969) has shown, by a study of the sex ratio among mature cattle in various parts of India, that in fact Indian rural society manages to eliminate a substantial proportion of an unwanted sex before maturity. The need for work oxen on the one hand and for milk and milk products on the other means that a large cattle population must be maintained if needs are to be met at the current low level of productivity. It is therefore probable that as things are at present the actual surplus cattle population is quite small, and unimportant in India's agricultural balance sheet. Revisiting India after 30 years, one can see an improvement in the standard of feeding, particularly of dairy cattle and milch buffaloes. Genetic improvement has so far been small, but crossbreeding programmes are in operation with dairy stock. At the same time, improvement in crop production leads to the introduction of tractors to supplement, and to some extent to replace, work oxen. Indeed, as India's soil and water conservation problems are tackled, tractors will increasingly replace oxen. The social impediments to the control of cattle populations will then have to be faced, as the choice will lie between low production by an uncontrolled number of underfed animals, and high production by genetically superior stock, adjusted in numbers to the resources of feed and forage that can be made available.

These are the technological considerations in the improvement of Indian agriculture, fertility level, water supply, and the performance of crop varieties and livestock strains. There remains the problem of the vast human communities dependent upon agriculture. The population of India is now about 540 million, and of these 70 per cent of the work force are engaged in agriculture. Undoubtedly the easiest way to feed these millions would be to reorganise Indian agriculture and displace a large proportion of those now engaged in peasant farming. Development has in all countries proceeded in this way, and the displaced rural population has either been absorbed into industry, as in western industrial revolutions, or has crowded into shanty towns on city margins, as is now happening throughout the developing world. India's own experience shows that the rate of growth of industrial employment is inadequate to absorb the people that would be released by such an agricultural reorganisation, and some other solution to the problem must be found.

The need to provide in agriculture, subsistence for all who have no other livelihood has led to the colonisation of the wastes in 14th century Europe, encroachment on the resting phase in shifting cultivation areas in Africa, and subdivision of small-holdings in present day India. It is the driving force behind the political demand in India for a ceiling on farm size, and for subdivision of large holdings to provide subsistence for the landless. As a political programme it is subject to a diversity of abuses. As a social force it is to be recognised and accepted as inevitable in present circumstances. Only by the provision of land can the rural workless be assured of subsistence. At the same time, however, the growing cities must be fed, and they will be fed not by subsistence holdings, but by farms large enough to yield a substantial surplus beyond subsistence, and hence integrated into the commercial economy. Thus India is faced with the need to pursue two incompatible policies at the same time. If industrialisation is to succeed, and standards of living are to be raised, the commercial sector of her agriculture must be encouraged, and must be enabled to produce the necessary surplus over home consumption to feed the growing cities. At the same time, until alternative livelihood is available, subsistence farms must be provided for all those who need them, by the subdivision of existing holdings.

The transfer of people from agriculture has rarely been achieved without hardship and distress. There is no adequate social and economic theory from which to derive the principles on which it should be planned. It is the most pressing need in India today. Redeployment will not come about by importing capital, or by the unplanned and unorganised growth of population in the great cities. If it is achieved it will be by the replanning of rural communities to increase their own contribution to their own needs. This is a philosophy of which Gandhi would have approved. It is involved in the political commitment of India to greater equality and to the elimination of poverty. Let there be no misunderstanding on this matter. Ways of doing this have not yet been devised, either in India or anywhere else in the world. It is a problem for research and it is worthy of the best men that India can find. In its solution India could lead the world.

PART IV: AGRICULTURAL STRATEGY

CHAPTER 10

AGRICULTURE AND DEVELOPMENT

Development is the process by which human communities satisfy three basic needs. First is the means of subsistence, and the emergence of man from hunting and food gathering to agriculture made subsistence secure and opened up the prospect of satisfying the second need. This is for a rising standard of living through the enjoyment of other things besides agricultural produce; the fruits of the intellect such as law and order, religion, learning and research, and the products of crafts and industries. Only recently, in communities that have attained affluence in respect of the second need, has the third become apparent. This is the need for a healthy and satisfying relation between man and his environment. The establishment of such a relationship involves conservation in all its aspects, and the achievement by man of control over his own numbers.

In this chapter the role of agriculture will be considered as the foundation on which the satisfaction of all these needs must depend. Development has gone on disproportionately in different parts of the world, and there has arisen a great disparity between rich and poor nations. In the last quarter of a century considerable efforts have been made to reduce this disparity, and some progress has been made in raising the standard of the developing economies. Nevertheless, the gap between rich and poor has continued to widen, and in recent years some disillusion with the progress of development has been apparent. If this process is to gain fresh impetus the place of agriculture in it must be more clearly understood, both historically and in relation to present needs and opportunities.

The goods and services that make life more than subsistence

are produced by men released from agriculture when farm productivity becomes high enough to supply a growing community by the work of a declining labour force. In the 19th century the British Industrial Revolution created a world community in which the agricultural transformation was undertaken differentially. The British agricultural work force was reduced to supply the labour to man the factories. Migrants to the newly opened lands of America and Oceania created a large agricultural force to feed the factory towns, and the populations of the Third World were brought in to supplement the temperate region foodstuffs of the newly opened lands with tropical sugar, oilseeds, beverages, textile fibres and rubber. Thus there arose in the developing countries economies based almost exclusively on agriculture.

The people of these countries participate in the goods and services of the affluent society only in so far as they can sell agricultural produce to finance the purchase of industrial goods. This long range exchange of agricultural produce for industrial goods has arisen in the last 150 years. It has been the means whereby many countries have taken the first steps in development, but it is inadequate to support a further stage in the process. It is the economic counterpart of the political system of imperialism and colonialism. The political system is obsolete and has passed away. The economic system persists and its continuance is the basis of the charge of neo-colonialism so often levelled against the great metropolitan powers.

The new temperate region communities did not remain long in this satellite position. They also industrialised, drawing off their agricultural work force to man their own factories. The United States has become an industrial super-power. Australia has a population that is predominantly city dwelling. Even Denmark, which has most successfully exploited this relationship, is now rapidly developing the urban sector of the economy. So the process of industrial development in an agricultural economy that began in Britain and spread to the United States, is now going on in all the temperate region agricultural economies.

The countries of the Third World must inevitably follow this road of balancing their economies. It is not possible to contemplate an expansion of the sale of cash crops adequate to finance the import of the industrial goods necessary to provide the rising standard of living to which they aspire. A rising standard can only come from a progressive increase in activities outside the agricultural sector by their own communities. This involves the transfer of labour from agricultural to industrial pursuits. Though it has gone on since civilisation began, it is rarely easy. Hill (1956), for example, in his *Tudor and Stuart Lincoln*, has given an account of an English town in the 16th and 17th centuries, embarrassed by migrants from the countryside. In pre-Industrial Revolution times, employment in the towns was not expanding, and the established town communities did all they could to protect themselves and their crafts from the flood of rural unemployed, which constituted a threat of cheap labour. The countryside, on the other hand, was producing a labour surplus for whom there was either no livelihood at all, or at best, bare subsistence as cottars squatting on the waste. The same situation is to be seen today in all the cities of the Third World. We have no answer to the shanty towns and unemployment of the urban fringes in developing countries except the development of industrial employment that will both absorb the productive capacity of these people, and create the demand for agricultural produce that will raise the living standards and increase the attractiveness of the rural areas.

The pattern of transfer is to be seen clearly in mediaeval England and in many developing countries today. In modern England it is obscured by the effects of long distance transport established in the Industrial Revolution. Urban employment opportunities in Britain have not only caught up with population growth and with the rate of transfer out of rural occupations; they have exceeded them to the extent that a substantial demand has grown up for immigrant labour. Unfortunately, the nature of this situation has not been appreciated and immigration has come to be regarded either as a means of meeting labour shortages, or as the creation of a race problem, according to the personal views of those concerned. It is in fact a response to the fundamental forces of development. Britain has established a system of exchange of industrial goods for agricultural produce over a much wider area than the British Isles. It is a natural consequence that there should be migration from the rural to the urban sectors over the whole of this wider area. Migration from the West Indies and Pakistan to the English Midlands is the modern equivalent of migration from rural Lincolnshire to Lincoln in Tudor and Stuart times.

Looked at in these terms, the rate of immigration into Britain is infinitesimal compared with the urban migration that would be necessary to achieve a steadily rising standard of living throughout the trading area. If the societies of the developing countries are to move towards a new balance between agricultural and industrial production, transfer of population out of agriculture can only take place on an adequate scale if it is planned within the countries concerned. To this Britain should contribute, both in the interests of development in her trading partners, and for the sake of good race relations within her own borders.

Labour for non-agricultural development is not difficult to obtain. Migration from country to town goes on at a rate in excess of the capacity of the urban sector to provide employment. This is partly due to lack of capital, and substantially also to inadequacy of training and educational resources. Unskilled labour direct from the rural sector is employable in only a limited range of urban occupations. Moreover, the absorption of even this amount of unskilled labour depends on the availability of an educated and skilled supervisory and administrative cadre to organise and direct urban enterprises.

Since the last war, education at all levels has been given high priority in developing countries. It has been regarded as a means of escape from the countryside and entry to, and advancement in, urban life. Administrators, planners and politicians alike have deplored the 'drift to the towns', but have been powerless to stop it. They have been right in so far as education is as important for the development of agriculture as it is for progress in the towns. It is impossible to provide a safe city milk supply, for instance, without on the one hand dairymen who can read the material provided by the Medical Officer of Health, and on the other, agricultural scientists who will breed productive cows, and will devise farming systems on which productive cows can be kept.

Nevertheless the urban migrants are right, in that agriculture is traditionally balanced to provide the food and other products needed by the community dependent on it, and expansion in agricultural production is only possible in response to expanding demand. This comes from the expansion of the non-agricultural sector of the community. So the countries of the Third World, following the lead of the successful Western World, develop industries in their cities, and watch helplessly while their rural peoples migrate to the urban centres in unmanageable numbers.

Modern industrial production, like modern agriculture, does not require a large labour force. As developed in the western world it is large scale, capital intensive, and extremely economical of labour. Even in western countries full employment has only been achieved recently, and even then must be regarded as counterbalanced by under-employment in the Third World sector of the trading area. In the industries of the Third World, large advances in production result in only small increases in employment.

In this respect, the record of industry compares unfavourably with that of agriculture. Agricultural communities have throughout history accepted the obligation to provide the means of subsistence to all their members. Hence, when the population presses on the land, holdings are subdivided, even to the point where the most industrious find it hard to gain subsistence. No such reallocation of urban resources goes on to meet the need of growing urban populations. Indeed, in the industrial sector of developing countries, the establishment of high wage rates for those who have gained urban employment, and the introduction of labour saving practices, go on while urban unemployment increases. The illusory attraction of high wages for a limited labour force draws even more people from the countryside, where social custom would at least have ensured them a subsistence.

These restrictive practices in urban employment, together with the more generous social customs in rural areas, conspire together to make agriculture the occupation of last resort. Authorities on the subject maintain that development must be predominantly development in agriculture. There is little consideration of what possible urban market there can be for the quantities of agricultural produce that would have to be sold to give the purchasing power on which a higher standard of living could be attained. Those concerned with land reform are all too often reduced to dividing up farms on which a substantial saleable surplus is produced, into peasant holdings adequate at best for little above subsistence. For the administrator there arises the dilemma, so acute in India at present, that the farmers with the larger holdings have the resources to respond to new technology and to generate the food supplies so urgently required, yet population pressure in the rural areas is such that subdivision is imperative, and state governments resort to imposing ceilings on the size of holding a man may farm. So, caught between population growth and the inelasticity of urban labour demand, the peasant is condemned indefinitely to little more than subsistence. Indeed, some have regarded the 'green revolution' as retrograde, or even as a calamity, because it has given the enterprising increased wealth and hence the opportunity to increase the size of their enterprise. Since this cannot be matched by the absorption of labour in non-agricultural pursuits, the poor in the country areas get poorer. Those who argue that development must be development in agriculture will do well to ponder the dilemma of the Indian situation. There is no future this way, and one is tempted to hope that the advocates of land reform may be persuaded to reorganise the urban sector as they would reorganise the land. For if standards of living are to rise, employment in the urban sector must increase.

It is not only at the level of the individual that the poor stay poor while the rich get richer. It is characteristic of states also, since industrial growth has been based on the exploitation of economies of scale. Mass production depends upon a mass market. Hence the need felt by British industry to gain entry to the Common Market, and hence also the difficulty experienced by African countries in attempting to establish their own fertiliser industries. For this reason the United States and Canada have achieved the super-state economy. Australia has the size to achieve it in the future if the population grows sufficiently. India, with a vast population and great resources, is already well on the way to massive economic growth, even though no corresponding growth in employment can be foreseen. New Zealand, the African countries and most Latin American countries are in a very different position. They do not have now, and have no hope of having in the future, populations large enough to sustain such economies. To the United States, the USSR, and Western Europe they are of marginal interest as agricultural producers. Britain, through whose industrial revolution their agricultural economies came into existence, is in process of cutting them loose and entering the emerging Western European super-state.

The urge to increase in size permeates modern society. Industries are merged to make vast combines, and farmers are advised to create large units in the interests of efficiency and economy. The advice so often heard in Australian agricultural economics – 'get big or get out' – pervades the whole of society. In agriculture the small farmer stays on because no answer is given to his question 'Where to ?'. In a real sense this is the question to which an answer must be given on a national scale also. All small economies are equally affected. The problems of New Zealand with her economy heavily dependent on agricultural exports are matched by those of Trinidad with the beginnings of an industrial economy based on oil. The great states are no more willing to accept the products of Trinidad's infant industries than they are the products of New Zealand's well established agriculture. But modern industrial thinking does not contemplate the establishment of balanced economies in countries with populations of only a few millions, so our current philosophy has no solution to offer to the problems of small states.

If this were the end of the matter, most of the work on development in recent years would have been to no purpose, but the economic achievements of the great states are neither satisfactory in themselves nor necessarily the only type of progress that could be planned. The first Industrial Revolution was based on steam power. In the absence of a public transport system and with the exploitation of large scale power supplies, industrial labour was concentrated in dense urban communities, within walking distance of the steam boilers that powered the factories. This was the origin of the modern industrial city and its direct descendant, the great conurbation. Previously industrialisation, as for instance the woollen industry of mediaeval Britain, had not resulted in draining the countryside of people and crowding them into towns. Nor need it do so now. With the electric grid and the ease of transport of the 20th century, urban concentration is no longer inevitable.

The conurbation is a bad habit, not an economic necessity.

It is inconvenient, inefficient, and expensive. It aggravates the problems of supply and transport, and of disposal. Even in well-watered countries such as Britain, the supply of an adequate quantity of pure water to the London, Birmingham and Manchester regions is a major, and costly, engineering enterprise. So also is the transport, both of work people and of supplies of food and raw materials. Further, the disposal of effluent and waste presents problems that have yet to be solved. The results of continuing uncontrolled growth are not to be contemplated. Yet in none of the great states has any system of control been devised. If this is development, the people of the small states may well conclude that they are fortunate to be excluded from it.

This, though the most recent, is not the only form of development, and it is well to remember that most of man's great achievements were accomplished by small communities living in small independent states. The greater part of our cultural heritage was generated in small city states where there was the intimate relation between town and countryside that facilitates the deployment of the work force according to social need. In such states arose the art, literature, and architecture of India, of Italy and of England. Tudor and Stuart England had no larger population than modern New Zealand, but it gave birth to Shakespeare, Marlowe, Milton, and the founders of the Royal Society of London. The Italian states that gave us Galileo, Copernicus and Dante were comparable in population and indeed in natural resources to at least the larger of the West Indian islands or of Mauritius. That so much has been done in small communities in the past gives hope for the future.

The problems with which the developing world is faced will be solved, not by the enlargement of world trade, but by the deliberate establishment of a balance between farm production, and crafts, industries and services, not only within states but also in the smaller communities of which they are composed. It is generally agreed that the population of these countries will inevitably remain country-based. The only alternative to a continuing subsistence economy is to organise for themselves something better within their own rural setting. It is not easy to develop an alternative to large-scale industrialisation. The power of the great industries is greater now than when the

Lancashire cotton industry crushed the textile crafts of India. Yet something is being done. In India today cottage industries have advantages in individuality and craftsmanship that are of importance in India's trade, and the Intermediate Technology Development Group based in Britain is engaged in fostering enterprises that can be mounted with little capital and modest skills. The philosophy of such an approach to development, and its consequences for the transfer of people from agricultural employment, have been considered by men like Gandhi in India and Nyerere in Tanzania. The Gandhian philosophy still motivates many people in India, and Nyerere's (1967) views on the structure of the society he wants to build, and the educational system necessary to achieve it, are wise and timely. It is from this kind of thinking that an alternative to Western industrialisation will emerge.

FEEDING THE PEOPLE

A man's first preoccupation is with his food supply, and only in so far as this is assured is he able to devote his attention to the satisfaction of other, secondary wants. It has been the great achievement of agriculture that its productivity has risen at a pace that has allowed a progressive increase in the size of the human population. In recent years the increase in productivity has been such as to permit a rate of increase that is generally described as an 'explosion'. This is a Malthusian situation in reverse, for the prevention by medicine of the misery of sickness would have been of no avail unless there had been an increase in food supply adequate to prevent the misery of starvation. In feeding this rapidly increasing population there is, rightly, considerable anxiety lest agricultural productivity fail to keep pace with population growth. A review of the resources available is therefore appropriate.

The dominant climatic factors in agricultural production are temperature and water supply. In no part of the world is it too hot for agricultural production. In Arctic regions and at high altitudes it is too cold. On account of the cold, a large area of northern Eurasia and northern America, all of Antarctica and small areas of high mountain throughout the world can be written off as agriculturally unproductive.

The availability of water involves more complex considerations, since both supply and consumptive use are involved. For example, 600 mm per annum in Cambridge, England, is adequate to support an intensive and highly productive agriculture, whereas in the 500–750 mm rainfall belt in Africa south of the Sahara, agriculture is at a low level of productivity and is regarded as distinctly precarious. Likewise in Australia, wheat is grown in areas with an average rainfall as low as 400 mm but the system is both extensive and subject to considerable risk. Below about 400 mm per annum only extensive pastoralism is practised, and this also is subject to hazards which result in wide fluctuations in stock populations, often by sheer starvation and thirst.

At the other extreme, agriculture is unproductive in areas where rainfall is substantially in excess of consumptive use. Consumptive use, or evapotranspiration from soil and vegetation, being a function of the solar energy received, increases with increasing temperature and with increasing solar radiation. It is thus greatest in hot dry areas and so a hot country with a 500 mm rainfall is semi-arid, whereas a cool temperate country with the same rainfall is mesophytic in ecology. In hot countries, consumptive use is less where there is a considerable cloud cover and a high rainfall than where there are clear skies and little rain. Thus it comes about that, in tropical countries, rainfall greater than about 1500 mm is considerably in excess of vegetation requirements, and gives rise to a surplus which is discharged either by run-off to rivers or by percolation to a water table. In such circumstances soils are impoverished either by erosion or by leaching, and carry either a heavy forest cover or a low grade grass savannah. In either case, the practice of agriculture is unrewarding, and except under such tree crop cultures as rubber and cacao, results in rapid further deterioration in soil fertility.

Thus it may be said that the world's most important agricultural lands are those with a rainfall between 500 and 1500 mm per annum, in cool temperate, temperate or tropical climates. To these may be added river valley lands with less than 500 mm of rain but with water from elsewhere that can be used for irrigation. From these productive lands must be deducted hill and mountain areas that are too steep for agricultural use, and areas described by D'Hoore (1968) as 'soil-less' areas, areas which do not support an effective vegetative cover, even in the wild.

These agricultural areas have not been uniformly colonised. The exploitation of temperate Europe and China and of warm temperate and tropical India has given support over the past 4000 years for the largest concentrations of human population. On the other hand, the agricultural exploitation of the moderate rainfall areas of Africa has been less successful and no large populations have arisen beyond the Nile valley and the north African littoral. These disparities are associated with variations in the third great environmental influence on agriculture, namely soil fertility. Soils are the result of the interaction of geology and climate over long periods of time. In the temperate regions of the northern hemisphere the alternation of Ice Ages and temperate climates has left great areas of soils of good intrinsic fertility, good depth, and a structural stability that has withstood long continual cultivation. In India also, the action of weathering on rich geological material has generated great areas of intrinsically good soils, both on the northern alluvium and in the peninsula. By contrast, in Africa and in parts of North and South America – the south eastern United States, Venezuela, Guiana and much of central Brazil – there are vast areas of ancient soils, weathered and leached to the point where the nutrient status is very low and soil structure is unstable and easily broken down. Australia also is a continent of extremely poor soils, and extensive 'soil-less' regions. Indeed, with its long history of arid and semi-arid climate, it is fair to say that over most of the continent soil forming processes have not gone on to any advanced stage.

Agriculture began, and farming peoples spread over the world, against a background of resources in land that appeared inexhaustible. That stage is long past in India and Britain, but it is recent history in North America, and is current experience in parts of Africa and Latin America, and in Australia. When land is in abundant supply it is not valued and is often overexploited. Fortunately, over most of the world man's agricultural activities, though they have changed the soil and vegetation, have not ruined it. Neither shifting cultivation nor the fertility transfer of settled farming involve a net loss of nutrients. Only when man has created large urban populations, and has had to establish waste disposal systems that end in the sea, has he created a real drain on the fertility resources of the land.

The impact of developing human societies on the farms on which they depend is best studied in simple land use systems. Principles so established may then be applied to the circumstances of modern communities, where the relation of society to the land it occupies involves very much larger areas. The classic study of a simple system is W. Allan's (1965) *The African Husbandman.* Allan has given an account of the relationship of shifting cultivators to the land of low fertility and unstable structure, that they farm in Zambia. He has traced the effect of population growth on land use and has shown how, with a growing population and a limited land area, the balance between production and demand is held by increasing the frequency of cropping at the expense of the rest period. Since the maintenance of the fertility level under shifting cultivation depends on the remobilisation of the small stock of nutrients by the regenerating natural vegetation, there comes a point in the progressive shortening of the rest where remobilisation is not adequate and a decline in fertility sets in. Allan's studies led him to propound the concept of a 'critical population density' which he defined as the density above which a progressive deterioration takes place in the land resources available to the population. The concept of the 'critical population density' is applicable to all forms of human society. This density is reached sooner or later where an expanding population depends upon a static agricultural system and a limited land area. When it is reached, degradation of natural resources goes on to the point of catastrophe for the human population, unless a more intensive system of agriculture can be devised or fresh land made available. Catastrophic decline in human populations of this kind has occurred more than once in history. Van Bath has ascribed the population decline in Europe in the 14th century to the overrunning of the resources in land, and demographic collapse in classical Greece occurred in a similar situation.

This is essentially the situation which Malthus predicted from his argument that the rate of human increase would exceed the rate of increase in agricultural production, but Malthus' interpretation is inadequate, in that this is only one half of a two-sided supply problem. In agricultural production, the margin between hunger and surfeit is narrow. With an expanding population on a limited land area, the risk of hunger can indeed be serious, but equally circumstances can, and do, arise in which food production outstrips population growth and conditions of surfeit occur.

Population increase over long periods of human history has been matched by increasing agricultural production. According to van Bath (1963), from the Norman conquest to the mid 19th century the population of England increased from one million to 16 million. During that period, there were from time to time small import and export balances of food, but for practical purposes England was self-sufficient. Only once, in the 14th century, did the increase in agricultural productivity fail to match the increase in population.

This period was followed by one in which population and food supply were unbalanced in the direction of surfeit. In the second half of the 19th century the opening up of the New World after the British Industrial Revolution led to an excess of food production over human requirements. There followed three quarters of a century of agricultural depression, during large parts of which depression spread to the industrial sector of the economy. Thus over exploitation of the land may lead to the degradation of natural resources and poverty and hunger on the one hand, and to overproduction and economic depression, and poverty in the midst of plenty on the other. Both are parts of the larger problem of the balance between man and his food supply, in which unmanageable surplus may occur as well as desperate insufficiency.

In the modern era, which began with the great increase in agricultural production in the newly opened countries of the Americas and Oceania, the farmers of the New World suffered in the long agricultural depression as did those of the Old. It was not that the Old World could not compete with the vigorous agriculture of the virgin lands of the New. It was that neither could do more than scrape a bare living when together they had drowned their own market. Moreover, the tropical countries were similarly affected. In this period the contribution of agriculture to tropical development was restricted to growing crops that were in demand in temperate regions, but were not in competition with temperate farm products. Where there was competition, great difficulties arose. The long continued troubles of the sugar industry arose from competition between temperate and tropical production in a period of surplus. Wheat, of which an export surplus developed in India following the extension of irrigation in the north west of the sub-continent, was in direct competition with wheat from North America. Only products such as tea, coffee, cotton, cacao and oilseeds offered export possibilities. Consequently, it was to these crops that agricultural departments in developing countries directed most of their attention.

For Africa, the exchange of tropical agricultural produce for the industrial goods of the West provided the beginnings of economic advance. It was not that food crop agriculture made no contribution to development. Food crop production was readily increased when there was a market for it. A case in point is the development of food crop farming in the Middle Belt of Nigeria and corresponding regions in Ghana to feed the cacao and oil palm farmers of the forest region. It was that there were locally no large urban populations producing industrial goods for exchange for food.

Such towns as there were in tropical Africa grew up to service the exchange of tropical cash crops for imported industrial goods. The food habits of the expatriates who organised the trade led to a demand for temperate region foods, and transport, storage and cooking practices adapted to urban needs had not been worked out for tropical food crops. Hence the food habits of urban Africa became dominated, both for reasons of prestige and on account of convenience and reliability of supply, on imported foodstuffs such as wheaten flour, and tinned meat and dairy produce.

The situation changed after the second world war. Food surpluses had disappeared and, with the exception of the Korean interlude, the markets for cash export crops were amply supplied. It became clear that expansion in developing economies could no longer be based on expansion in the production of crops for export. Local industries were fostered, local towns began to expand, and more attention was paid to food agriculture.

With the beginnings of industrial development and the accompanying urban growth in the 1950s, attempts were made to 'modernise' local food farming. In Uganda, for example, when the Owen Falls dam was built to supply power for urban development, the Government set out to ensure the food supply for the large labour force assembled at the dam site by opening a large mechanised food farm in South Busoga. It was a costly failure. Local agriculture expanded without difficulty to meet the need for local foodstuffs, and supplies for engineering and managerial staffs were in any case imported, either from the highlands of Kenya, or from overseas.

The urban growth that followed has stimulated food farming as a natural extension of indigenous farming practice. Farming fully integrated in the cash economy has grown up in southern Uganda within easy transport range of Kampala and Jinja. Moreover, production is not confined to traditional African foods for town labour. The sector of the food industry formerly dominated by the import trade is now supplied in part. Ample supplies of fruit and vegetables are produced locally, meat and poultry supplies are substantial. Dairying is growing rapidly, and as the numbers of productive dairy cows increase, local production will take over more of the local market.

The history of the development of food farming in Uganda illustrates the course of development that is going on in tropical African countries. The stage reached varies from country to country. Uganda is favourably situated, with good farming country near the centres of urban development. Ghana may be taken to illustrate the consequences of a less favourable agricultural geography. Urban development is concentrated in the coastal ports, to serve the export trade in cacao. The coastal strip carries a degraded savannah vegetation, and is regarded as agriculturally unattractive. North of the savannah is the forest belt, in which the cacao is grown. Local foodstuffs, which are chiefly root crops, are grown partly in the forest and partly in the orchard bush transitional zone to the north. Beyond the transitional zone is the true savannah, where sorghum and bulrush millet are grown and cattle are kept. Thus the urban concentrations are a long way from the best food producing zones. Some dairying and poultry production has been started on the coastal savannah, but even this depends initially on imported concentrates.

These two examples illustrate the opportunities and the problems that face African agriculture in the later decades of the 20th century. Industrialisation and urbanisation have begun, and the opportunity now is for agriculture to feed the growing towns. Producing enough food is not difficult, as there is still land to spare. What is needed is food within transport range of the towns, of kinds and qualities suited to an urban diet, and this calls for imaginative research and development between the farm gate and the kitchen sink as well as for the intensification of crop and animal production. Transport and storage, a food industry producing convenience foods, and processing and cooking to make local foodstuffs suitable for restaurant eating, office meals and easily prepared home meals, all require study. This development is necessary not only to enlarge the opportunities for local agriculture, but also to reduce the use of foreign exchange earned by export crops to finance food imports. Countries like Uganda have made substantial progress on these lines. Countries like Ghana that are geographically less favoured have not gone so far. Both need to reassess the agricultural potential of the farm lands on which the towns will depend. For food farming for the town market is intensive agriculture, and fertility levels much higher than those natural to African soils must be established if it is to be successful.

India, which was economically and socially more advanced than Africa at the beginning of the Imperial era, did not develop in the same way in response to the challenge of the Industrial Revolution. The invasion of India's markets by Western industrial goods was at the expense of India's indigenous crafts and industries. The export of indigo, jute, cotton, wheat and tea was an inadequate compensation for the ruin of the exchange of farm produce for city goods on which the Indian economy had long been based. The balance was only redressed by the establishment in India of western type industries and by tariff protection for them.

India entered the post-war period with a substantial measure of industrialisation achieved. In the post-partition and independence period industrialisation and urban growth have gone on, and India is now an industrial nation. India is poor, but the country is not a 'developing' nation in the sense that African countries are. In consequence the agricultural problems involved are very different. Population increase has been very rapid, and new land was limited to areas such as Rajasthan where major irrigation schemes were possible, the Terai where malaria control opened up jungle land for farming, and Dandakaranya, together with small extensions all over India by irrigation, clearing of perennial weed infestation and some forest clearance. The main emphasis, however, was of necessity on raising the productivity of land already in use, and this meant a major fertiliser programme. This was highly effective on irrigated land, and the increased productivity of the irrigated land has been the main factor in the doubling of India's food grain production between 1950 and 1970.

Some underestimate of the increase in food production necessary to match population growth, together with two bad monsoons in succession, led to serious food shortages in the late 1960s, but the position was better than it appeared. Indian agriculture is diversified, and produces all the elements of a satisfactory diet. Town food supply is an enterprise in which Indian farming has been engaged for several millennia. Convenience foods and restaurant eating are new phenomena, but there is no need to resort to imports to meet them. The problem in India is not quality but quantity. Increasing quantity can be foreseen for some time to come, and when the means of raising fertility levels on rainfed lands have been worked out, vast further increases will be possible. The intractable problem is poverty. It may well be that the drive to match the growing population with an adequate supply of food will be frustrated by the failure of the social system to provide the hungry with an income with which to buy.

In temperate regions the great expansion of agriculture into empty lands was checked, first by drought and wind erosion in the 1930s in the United States, and then by the demands of the second world war. After the war improved technology led to higher productivity. In Britain the need for imports of food was moderated, and in the United States surpluses arose, in spite of acreage restriction. The index of farm output in the United Kingdom doubled between 1939 and 1964 (Fig. 8, p. 77), and in the United States, rose from a figure of 59 in 1920 to one of 116 in 1965 (Statistical Reporting Service, USDA, 1966). However, the continued abundance of food stimulated the growth of urban occupations and massive migration from the land to the cities. In the United Kingdom there were 800 000 agricultural workers in 1939, nearly a million in 1947, and less than 600 000 in 1964 (Fig. 8, p. 77). In the United States the rural farm population fell from a peak of 32 million in the decade 1910–20, when it made up 30 per cent of the population, to 12·4 million in 1965, which was only 6·4 per cent of the much larger total population.

Urban expansion has had only local effects on land use patterns in great countries such as those of North America and Australia. In Western Europe, on the other hand, competition for land is now severe. Resources in land in Britain, for example, are small in proportion to the population, and good farm land has progressively been given up for urban growth. Ten per cent of the farmland of England and Wales, 1 200 000 hectares, has been lost to farming in 60 years, and the current rate of loss is 20000 hectares per year.

The areas nearest the town markets have advantages for agriculture and particularly for horticulture, but as the town grows this is the land it spreads over. The cities have developed in good agricultural areas because these were areas where the resources for urban development were available. Moreover, the stimulus of urban development led to the intensification of the farming round about, as shown by van Bath for the Netherlands, and as can be seen in modern Britain. Lancashire, with its great city population and rapid urban growth, is the county with the highest agricultural productivity in Britain.

While suffering these losses of the most profitable land near the towns, British agriculture has retreated somewhat from the marginal lands also. This has gone on by afforestation, by the abandonment of the higher farmsteads, and by running the land together in more extensive units. Amalgamation and withdrawal would have gone further but for the special support for hill farming. This was given as aid to an impoverished sector of the farming community, but it has had a wider impact than simply farm support. The hill areas are already, and will become increasingly, recreational areas for Britain's vast urban population. To be attractive and satisfying, a recreation area must not only offer direct holiday services – good access, hotels, garages and sports facilities – but must support a stable rural community, practising good land management. In this respect support for agriculture and indeed for forestry, contributes to the maintenance of conditions under which recreational activities flourish.

The problems of Australia are of a different kind. Australia has been developed as a supplier of agricultural produce to Britain, and it is threatened by the contraction of the market for its agricultural produce following the increase in productivity of the good lands of Britain and Western Europe. Australia has gone far in the establishment of her own internally balanced market. Her industries have been developed and protected and a large proportion of her population is city dwelling. The creation of a balance, however, calls for restriction of the expansion of agriculture as well as for encouragement of the expansion of industry. Controls have from time to time been imposed on the production and marketing of Australian agricultural produce to meet marketing difficulties. With the expectation that world food supplies will be more adequate in the next few years than they have been in the recent past, the need has arisen to plan a balance between the agricultural and industrial sectors of the economy. In present circumstances the increases in agricultural production that will be required can readily be gained by the improvement of good lands near the great centres of population, where the exchange of agricultural produce for urban goods can be kept in balance by migration of people between town and country. The expansion of agriculture by the clearance of new land and the bringing in of unproductive cleared land by minor element amelioration is unnecessary, and for some time to come is likely to be unprofitable.

The enterprise of feeding the world population can now be reviewed. Agriculture has met human need extraordinarily well, but there have been periods of unbalance, both of excess and of deficiency. The great increase of the late 19th century came from the farming of new lands. The shortages of the post-war period culminating in the serious deficiencies of the middle 1960s were generated by success in preventive medicine leading to explosive population increase. No major famine occurred because North American production was sufficient to supply the needs of the deficit countries of Asia, and modern transport and distribution systems made it possible to deliver the food to the areas of need. The advances that have filled the food gap of the middle 1960s were due to the improvement of productivity of land already in cultivation. So the pendulum swings as medicine, agriculture and engineering alter the balance of demand, supply and distribution.

The establishment of a stable balance between need for food and agricultural production now depends much more on the integration of industry and farming on good lands in areas of large human communities, and much less on the extension of agriculture to vacant lands far from centres of population. Supply and demand for food is a local matter. For less than a century the special circumstances of Britain's industrial growth and the opening of the empty lands of America and Oceania made possible an international trade in food on which the economies of America, Argentina, Australia and New Zealand were developed. But even for Britain there are limits to such a system of exchange. A strong and productive home agriculture is necessary to maintain economic stability, and the wider economic field of the European Economic Community is planned on the basis that its members will depend for their food on the good lands of Europe, and not on the poor lands of Africa or Australia. In developing countries, with very limited resources on which to trade on the world market, dependence on other countries for anything beyond a small proportion of the food supply is out of the question. This is illustrated particularly by the experience of India and Pakistan in the past decade when the shortfall in home production had to be made good by imports. Only by gifts and concessionary purchases was it possible to meet the need.

The demand for food is inelastic as well as local, and the future of agriculture will depend on carefully planned matching of local supplies to local needs. Thus, the United States may well again experience the problems of the management of surpluses unless the expansion of the American urban economy results in a further substantial reduction in the numbers engaged in agricultural production. Indeed, the risk of this occurring is increased in so far as deficit countries such as India and Pakistan exploit the results of agricultural research and balance their needs with their own supplies. Further, those developing countries that still have unused resources in land will best foster their own development by putting a high priority on substituting locally produced foodstuffs for those now imported and thereby releasing foreign exchange for vital imports.

Finally, in whatever economic system he works, the farmer will remain the provider of the means of life. On his efficiency depends the adequacy, and the cost, of food. For some time to come he can provide enough. It is probable that he will provide it cheaply. Cheap food means poor farmers. Farmers the world over are poorer than their urban cousins. They are likely to remain so for two reasons. First, agricultural production is so hazardous and human need for food so inelastic that no community would be secure unless its farmers budgeted for a modest surplus. So farm prices will tend to be low so long as agriculture operates in the situation of surplus that is necessary

9-2

to meet human need. Second, and more important, farming is, and will remain, a way of life. In urban society, economic activities and the way of life have been ruthlessly separated. A man works at a distance from his home and his family. He earns money in one place by strictly economic activities in order that he may spend it in another according to criteria in which economics play a minimal part. The farmer is more fortunate. His economic activities and his way of life are not, and cannot be, separated. Put at its simplest, the family and the farm are one integrated unit, and this is something of enormous human value. In recent years it has often been said that an advanced society can no longer afford an industry conducted as a way of life. Yet if the farmer, with purely economic motives, removes his hedges and uproots his parkland trees, he is accused of destroying the amenity of the environment that he holds in trust. Too often his critics want it both ways.

Fortunately, a regard for the environment is increasingly a part of our social thinking, and it is to be hoped that this will carry with it an appreciation of the part that farming as a way of life has played in the preservation of a good rural environment, and the extent to which the future of the countryside depends on this precious feature of agriculture. So it is right to acknowledge that joint concern with, and commitment to, the farm that only the farming family really knows. For this, there are many people still who are prepared to live with less of this world's goods than they could acquire in an urban occupation.

UNITED KINGDOM

METHOD OF COMPUTING THE LEVEL OF SUPPORT FOR MAJOR GROUPS OF AGRICULTURAL PRODUCE

In the *Annual Review and Determination of Guarantees* published each year (HMSO), the following data are given for the main groups of crops and stock:

(1) value of farm sales (including deficiency payments),

(2) cost of implementation of price guarantees (deficiency payments),

(3) cost of the relevant production grants and subsidies.

Support was estimated for six main heads of produce:

(*a*) milk and milk products,

(*b*) fatstock and wool,

(*c*) eggs and poultry,

(*d*) cereals,

(*e*) other farm crops,

(*f*) miscellaneous and valuation changes.

All sums relating to horticulture were deleted.

The unsupported value of the produce under each head was determined by deducting the relevant sum for deficiency payments from the appropriate value of farm sales. The cost of support under that head was estimated by adding to the cost of deficiency payments, a sum obtained by allocating the cost of production grants and subsidies in the proportions set out in Table A below. Then the level of support as a percentage of the unsupported value is:

$$\frac{100 \text{ (deficiency payments} + \text{allocation from Table A)}}{\text{unsupported value}}.$$

TABLE A. *Percentage allocation of grants and subsidies between the main heads of agricultural produce*

Grant or subsidy	Milk, etc.	Fat-stock and wool	Eggs and poultry	Cereals	Other crops	Miscel-laneous	Horti-culture (omitted)
Fertilisers	30	20	—	30	10	—	10
Lime	25	25	—	40	10	—	—
Ploughing up grants	—	—	—	75	25	—	—
Beans	—	—	—	—	100	—	—
Drainage	25	25	—	40	10	—	—
Calf subsidy	—	100	—	—	—	—	—
Hill stock and forage	—	100	—	—	—	—	—
Small farms and farm business	20	20	20	15	5	20	—
Crofting	—	—	—	—	—	100	—
Water supplies	100	—	—	—	—	—	—
Stock rearing and hill land	—	100	—	—	—	—	—
Farm improvements, farm structure and investment	20	20	20	15	5	20	—
Other	—	—	—	—	—	100	—
Administration	—	—	—	—	—	100	—
Northern Ireland	—	—	—	—	—	100	—

REFERENCES

Allan, W. (1965), *The African Husbandman*, Oliver and Boyd, Edinburgh.
Allchin, F. R. (1969), in Ucko, P. J. and Dimbleby, G. W., *The domestication and exploitation of plants and animals*, Duckworth.
Anon. (1956), *African Land Development in Kenya, 1946–55,* Min. Agr. Nairobi.
Anon. (1966), *The Common Market and the United Kingdom*, Westminster Bank, London.
van Bath, B. H. S. (1963), *The Agrarian History of Western Europe, A.D. 500–1800*, Edward Arnold, London.
Bunting, A. H. (1961), *Geography*, **46**, 283.
Clark, J. G. D. (1945), *Antiquity*, **19**, 57.
(1965), *Proc. Prehist. Soc.*, **31**, 58.
D'Hoore, J. L. (1968), in Moss, R. P. (ed.), *The Soil Resources of Tropical Africa*, Cambridge University Press.
Dodds, K. S. (1943), *Emp. J. Exp. Agric.* **11**, 89.
Ernle, Lord (1919), *English Farming, Past and Present*, Longmans Green.
Fisher, R. A. (1929), *The Realist*, **1**, 45.
Glover, J. and Robinson, P. (1953), *J. Agric. Sci.* **43**, 275.
Godwin, H. (1965), in Hutchinson, J. B. (ed.), *Essays in Crop Plant Evolution*, Cambridge University Press.
Gregory, S. (1957), *Quart. J. Roy. Met. Soc.* **83**, 543.
Gulati, A. M. and Turner, A. J. (1928), *Ind. Cent. Cotton Committee, Tech. Lab. Bull.* **17**.
Helbaek, H. (1966), *Palestine Exploration Quarterly*.
Hill, J. W. F. (1956), *Tudor and Stuart Lincoln*, Cambridge University Press.
Holliday, R. (1962), *Agric. Progress*, **38**, 88.
Iverson, J. (1941) (reproduced 1964), Land Occupation in Denmark's Stone Age. *Geological Survey of Denmark*, Ser. 2, Copenhagen.
Joblin, A. D. H. (1966), *E.A. Agric. and For. J.* **31**, 368.
Knight, R. L. (1946), *Emp. J. Exp. Agric.* **14**, 153.
Kohli, S. P. (1969), *Ind. J. Genet. and Pl. Br.* **29**, 24.
Lawes, D. A. (1968), *Cotton Gr. Rev.* **45**, 1.
MacNeish, R. S. (1962), *2nd Annual Report of the Tehuacan Archaeological-Botanical Project*, Robert S. Peabody Foundation.
(1964), *Science*, **143**, 531.
Manning, H. L. (1950), *J. Agric. Sci.* **40**, 169.
(1955), *Proc. Roy. Soc. B* **144**, 460.
Ministry of Agriculture, Fisheries and Food (1968), *A Century of Agricultural Statistics. Gt. Britain, 1866–1966*, HMSO.
(Annually) *Agricultural Statistics*, HMSO.
(Annually) *Annual Review and Determination of Guarantees*, HMSO.
See also, *Annual Abstract of Statistics*, HMSO.

REFERENCES

Nyerere, J. K. (1967), *Education for Self-Reliance*, Dar-es-Salaam.

Ogura, T. (1963), *Agricultural Development in Modern Japan*, Fuji Publishing Co., Tokyo.

Orr, John Boyd (1937), *Food, Health and Income*, Macmillan.

Parnell, F. R., King, H. E. and Ruston, D. F. (1949), *Bull. Ent. Res.* **39**, 539.

Penman, H. L. (1948), *Proc. Roy. Soc. A* **193**, 120.

Raj, K. N. (1969), *Investment in Livestock in Agrarian Economies*, Centre for Advanced Studies, Dept. of Economics, Univ. Delhi.

Rayns, F. (1961), *A Revolution in Arable Farming*, 3rd Lord Hastings Mem. Lecture, Norfolk Agric. Sta., Jarrold and Sons, Norwich.

Renfrew, J. (1967), *Thessalika*, **5**, 21.

Ryder, M. R. (1969), in Ucko, P. J. and Dimbleby, G. W., *The domestication and exploitation of plants and animals*, Duckworth.

Statistical Reporting Service, USDA (1966), *Agricultural Handbook*, no. 318.

Swaminathan, M. S., *et al.* (1970), *A New Technology for Dry Land Farming*, IARI, New Delhi.

Vishnu-Mittre (1968), *Trans. Bose Res. Inst.* **31**, 87.

Zohary, D. (1969), in Ucko, P. J. and Dimbleby, G. W., *The domestication and exploitation of plants and animals*, Duckworth.

INDEX

afforestation, Britain, 133

Africa, 88-91, 122; cash crops in, 91-4, 97, 128, 129; crop plants from, 5, 104; fertilisers in, 60, 93, 96-7; food crops in, 94, 95, 97-101, 128-31; livestock in, 32-3, 88-9, 98-9; soils in, 22, 23, 88, 95-6, 104, 126

agricultural advisory services: Britain, 79-80; India, 105

agricultural holdings: numbers of, United Kingdom (1935-64), 77, 78; pressure for sub-division of, in developing countries, 112, 119, 120

alluvial soils, 19, 21-2, 96; in India, 103, 107-8

alternate husbandry, in Uganda, 44, 45

America, 55, 56, 57; see also Canada, United States

antelopes, possibility of domesticating, 29

Arabia, 22, 55

artificial insemination, 31, 34, 84

Australia, 133-4; forage plants in, 28; industrialisation in, 85, 116, 120, 133; livestock in, 33, 53; rainfall in, 22, 124; soils in, 22, 52-3, 126

Baluchistan, 5

bananas, 19, 25, 26, 36, 91; in diet, 63, 64; improved varieties of, 38-9

barley, 2, 3, 5, 29, 30; yield of, England (1933-69), 50

beans, 30, 93; see also Dolichos, Phaseolus

beet, domestication of, 4; see also sugar-beet

bone meal, as fertiliser, 49

Brassicas, 4, 30

Brazil, soils in, 126

Breckland, East Anglia: intermittent cultivation of, 23, 43

Britain, 4, 8; agricultural depression in, 35, 57, 67, 81; agricultural revolution in (1950 onwards), 50-2, 61, 67-80; amount of fertilisers used in, 58-9, 77, 78; flow of food imports to, 55-6, 57; forests in, 13, 23, 133; Industrial Revolution in, 8, 55; 'New Husbandry' in (18th century), 7-8, 10, 11, 31-2, 48-9; output of agriculture in, 61, 77, 78, 132; soils in, 47; state support for agriculture in, 78, 79, 80-5, 137-8; see also England and Wales

buffalo (water), domestication of, 6, 104

cacao, 19, 24, 36, 38; as cash crop, 91, 129

Cajanus (red gram), 65, 102, 111

camel, domestication of, 104

cameloids (South America), domestication of, 6

Canada, industrialisation in, 85, 120

capital: for agricultural development, 8-9, 27, 49, 51; for fertiliser industry, 59, 96; for irrigation and plantation crops, 60

cash crops, 8, 35-6; distinction between food crops and, disappears with urbanisation, 94; world sufficiency of, 95, 97, 116; see also under Africa

cassava (manioc): in Africa, 91, 92, 93; in diet, 63, 64

cattle: in Africa, 32-3, 88-9, 99; in America, 57; in Australia, 33, 53; domestication of, 3, 6, 104; improved by both breeding and feeding, 31, 32, 33; in India, 111; in New Husbandry, 7, 48-9; numbers of, England and Wales (1935-70), 68, 69, 70, 72-3, 75; tropical breeds of, 34, 99

cereals: in diet, 63, 64; domestication of, 2, 5, 103; improved varieties of, 29, 40-1; increasing production of, India, 105, 110, 131; state support for production of, Britain, 68, 82, 83, 84; see also individual cereals

Chenopodium, as early food crop, 6, 28

Cicer (Bengal gram), 102

clay soils, 21, 23, 25, 27; in Britain, 51, 69; in India, 103, 109, 110

INDEX

Ghana, 13, 92, 130, 131
goats, 26, 31; domestication of, 2, 3, 6, 104
granites, soils from, 23
grassland, 3, 13, 16; permanent and temporary, England and Wales (1935–70), 68–9; in tropics, 20, 44–5
Greece, 22, 25–6; early agriculture in, 2, 4
groundnuts, 6, 15, 104
guano, 58–9
Guiana, soils in, 126
guinea pig, domestication of, 6

Harappan civilisation, 5
herbicides, 50, 60, 78; and ecological balance, 11, 61, 86
Hevea (rubber), 28, 36, 92
hill land, Britain, 19, 76, 133
Hordeum, wild species of, 2, 3, 29; *see also* barley
horses: domestication of, 4; replaced by tractors in Britain, 57, 61
horticulture, 98, 130, 133
Howard, Sir A., 9, 52
husbandry, 43–54; in introduction of new cereals in India, 106–7; 'New', *see under* Britain

ignorance, as cause of malnutrition, 63, 94, 98
India, ix, 54, 102–13, 120, 131–2; cereal shortages in, 105, 134, 135; cereal yields in, 46; cottage industries in, 123; early agriculture in, 5, 55, 104; fertilisers in, 39, 52, 59–60, 96, 105, 131; 'green revolution' in, 52, 59, 63–4, 109, 120; improved crop varieties in, 39–41, 52, 106–7; livestock in, 104, 111; soils in, 9, 47, 52, 103, 104, 105, 107, 126
Indian Council of Agricultural Research, 42
Industrial Revolution: and agriculture, 8, 55; and conurbations, 121; opening up of new lands by transport systems of, 55, 81, 128; in United States, 85
industrialisation, 116–18; in Africa, 130; does not now require urbanisation, 121–3; in India, 112–13, 131

industry: inputs into agriculture from, 50, 51, 55–6, 57, 58, 67, 78, 105; integration of agriculture and, 7, 8, 67, 134; labour flow from agriculture to, *see under* labour; support for agriculture and, compared (Britain), 85
Intermediate Technology Development Group, Britain, 123
Iran, early agriculture in, 2
Iraq, early agriculture in, 2, 3, 55
irrigation, 7, 18, 20, 55; in India, 39, 52, 55, 102, 103, 108, 109, 131
Italy, early agriculture in, 4

Jamaica, 25
jungle fowl, domestication of, 6, 104

Kenya, 89, 92, 93, 99

laboratory animals, domestication of, 28–9
labour: in agriculture, United Kingdom (1935–64), 77, 78; for cash crops, Africa, 91; exported from areas of poor soils, Africa, 90, 93; flow of, from agriculture to industry, ix, 67, 112–13, 116–19, 132; supply of, and mechanisation of agriculture, 58
Lancashire: agricultural productivity in, 133; livestock in (1935–70), 72, 73, 74, 75, 76, 77; tillage in (1935–70), 71, 72
land: with agricultural potential, 25; in excess of demand, 49, 126; lost from agriculture, United Kingdom (1935–64), 69, 78, 132–3
land tenure, in Africa, 93
Lathyrus sativus, breeding of less toxic varieties of, 111
leaching of soils, 18–19, 20–1, 22–3, 125
leguminous plants: in Africa, 91; in Australia, 22, 59; in crop rotation of New Husbandry, 48, 49; in diet, 64, 65; domestication of, 2, 3, 7, 104; improved varieties of, 65, 111; in India, 63–4, 110–11
lentils, 6
liming of soils, 27, 49

143

INDEX

Lincoln (Holland division): livestock in (1935–70), 72, 73, 74, 75, 76; tillage in (1935–70), 71, 72

linseed, 6

livestock, 1–2; breeding and feeding in improvement of, 31–5, 51, 98–9; drugs and sprays for, 61; integration of, into farming system, 10, 94–5; intensive rearing of, 11, 51, 84; in New Husbandry, 7, 31–2, 48–9; numbers of, England and Wales (1935–70), 69–70, 72–7

machinery, agricultural, 57, 78–9; see also tractors

maize, 3, 30; in Africa, 92, 93; in India, 104

malaria control, new land opened up by, 108, 131

Malthus, T. R., 127

manure: in Africa, 95; in husbandry of Netherlands (16th and 17th centuries), 46; in New Husbandry, Britain, 7, 48, 49; now becoming a problem in effluent disposal, 11, 51, 84; used as fuel in India, 52

marling of soils, 49

meat: breeding cattle for, 32, 34; as source of protein, 3, 63; state support for production of, Britain, 82, 83, 84; in Uganda, 130

medicine, preventive: and population growth, 8, 32, 67, 105, 124, 134

Mexico, early agriculture in, 3, 6–7

migration: into Britain, 117–18; into towns, 118–19

milk, 33–4, 56; state support for production of, Britain, 68, 83, 84

Milk Marketing Board, Britain, 84

millets, in Africa, 90; see also Eleusine, Panicum, Pennisetum

minerals, in nutrition, 62, 65, 98

mixed farming: in Africa, 89, 94–5; tendency to abandon, in Britain, 10–11, 51

Montgomeryshire: livestock in (1935–70), 72, 73, 74, 75–6; tillage in (1935–70), 71, 72

Namulonge, Uganda: Cotton Research Station at, 21, 44

National Agricultural Advisory Service, Britain, 79–80

National Seed Corporation, India, 107

neocolonialism, 116

Netherlands, husbandry in (16th and 17th centuries), 46–7, 48

New Zealand, 55, 85, 86

Nigeria, water control in, 9

nitrogen fixation: industrial, 58; natural processes of, 48, 59

nitrogenous fertilisers, 58–9, 60; amount used, United Kingdom (1935–64), 60, 77, 78

nomadic herding, 88–9

Norfolk: farming in, 7, 10; livestock in (1935–70), 72, 73, 74, 75, 76; tillage in (1935–70), 71

nutrition, human, 62–5, 81, 94; urbanisation and, 98, 100

oak scrub, Greece, 26

oats, domestication of, 2, 4, 30

oil palm, 19; as cash crop, 91, 129

output of agriculture: United Kingdom (1935–64), 61, 77, 78, 132; United States, 132

Oxalis sp., as early food crop, 6

Pakistan, ix, 55, 135

Palestine, early agriculture in, 2, 3

Panicum (millet), domestication of, 4

peas, 5, 91

peaty soils, 27, 51

Pennisetum (bulrush or pearl millet, bajra), domestication of, 5, 6

pest resistance: breeding for, 37, 38; cropping practice and, 42; of newly introduced varieties, 41

pesticides, 50, 60, 78, 93; and ecological balance, 11, 61, 86

pH of soil, 24

Phaseolus spp. (beans), 3, 6, 65, 91

phosphates: deposits of, Uganda, 96; in fertilisers, 59, 60; soils deficient in, 51, 59

pigs: domestication of, 4, 104; numbers of, England and Wales (1935–70), 69, 70, 75; Swedish landrace breed of, 32

pollen diagrams, showing clearing of forest, 4

pollution of water, 86

INDEX

Polygonum, as early food crop, 28

polyploidy, in crop plants, 30–1, 37

population: critical density of, 127; percentage of, in agriculture, (Britain) ix, (India) 112, (United States) 85, 132

population growth: in Africa, 100; agricultural productivity usually matched to, 80, 81, 124, 127, 134; in England, 127; in India, 105, 120, 131; preventive medicine and, 8, 32, 67, 105, 124, 135; in Tehuacan valley, Mexico, 6–7

potash, in fertilisers, 59, 60

potatoes, 6, 30, 68, 104; control of acreage of, Britain, 84

poultry, 34–5; in Africa, 97, 99–100, 130; domestication of, 4, 104; numbers of, England and Wales (1935–70), 69, 70; state support for production of, Britain, 83, 84

poverty, as cause of malnutrition, 63, 81, 98, 132

productivity of agriculture: breeding and feeding of livestock and crops in improvement of, 31–42, 51, 99, 106; recent increases in, ix, 10, 124; usually matched to population growth, 80, 81, 124, 127, 134

protein in diet, 3, 33, 62, 63, 64; in Africa, 94; in India, 110

pulses, *see* leguminous plants

quality: breeding for, in cottons, 37; in food, a matter of local preference, 41; of new wheat and rice varieties, 107

railways, 55, 56

rainfall: and agriculture, 14–16, 18–20, 52, 71, 124–5; and erosion, 21; in India, 105, 109

rice, 19, 20; domestication of, 5, 6, 104; improved varieties of, 30, 40–1; in India, 103, 106, 107, 108; mechanised cultivation of, 90; protein content of, 110

river floods, prediction and exploitation of, 55

Romans, practice of manuring by, 48

root crops, 3, 19; in diet, 63, 64

rotation of crops: in New Husbandry, Britain, 7, 48, 49

rubber, 19, 28, 36, 92

rye, domestication of, 4

sandy soils, 23, 27, 89

seed supply, in India, 107

seed/yield ratio, for wheat (13th to 19th centuries), 45

sesamum, 5, 30

sheep: in Australia, 53; domestication of, 2, 3, 6, 104; Finnish landrace breed of, 32; in New Husbandry, 7, 48, 49; numbers of, England and Wales (1935–70), 68, 69, 70, 74, 75–6; wool of, 31, 83

shifting (cut and burn) cultivation, 4, 9, 19, 43–4, 95; in Africa, 22, 88, 89–90; in Zambia, 87, 126–7

sisal, 92

soil fertility: conservation of, 7, 9–10, 11, 52, 53; improvement of, 49; in New Husbandry, 24, 49; transfer of, 46–7, 126; varieties of crop plants responsive to, 40–1

soils, ix, 14, 20–7, 125–6; domestication of, 24, 27; erosion of, *see* erosion; fertility of, *see* soil fertility; improvement of, 4, 46, 49, 53; leaching of, *see* leaching; storage of water in, 20–1, 103, 107; structure of, 11, 53, 54; *see also under* Africa, Australia, India

Sorghum (dura, jowar), 37, 90; domestication of, 5, 6

specialisation, in farming practice, 51

steamships, 55, 56

stones, clearance of, 27

subsistence farming, in Africa, 90, 91

Sudan: irrigation in, 60; nomadic cattle herding in, 88–9; shifting cultivation in, 43, 90; soils in, 21, 25, 55

sugar, price of, 83

sugar-beet, control of acreage of; Britain, 83; yield of, England (1935–69), 50

sugar-cane: as cash crop, 91, 92; excess production of, 20, 128; improved varieties of, 30, 36–7

sulphur, in fertilisers, 96

swamp land, 16–17, 27; rice in, 19, 20

sweet potatoes: in Africa, 91, 92; in diet, 63, 64

INDEX

Tanganyika, groundnut scheme in, 15

Tanzania, soils in, 23

tea, in Africa, 92

technology: application of, to agriculture, 55–61, 78–9

Tehuacan valley, Mexico: agriculture in, 3, 6–7

temperature, and crops, 14, 102, 124

tillage: acreage of, as percentage of total crops and grass, England and Wales (1933–70), 71–2

tobacco, as cash crop, 91

tomato, domestication of, 6

tractors, 10; in Britain, 57, 61, 77, 78, 79; in developing countries, 58; in India, 111

transport, in development of agriculture, 56

tree (plantation) crops, in tropics, 19, 56

Trinidad: cacao in, 24, 38; infant industries of, 121

Triticum, wild species of, 2, 3, 29; *see also* wheat

Tropaeolum, as early food crop, 6

trypanosomiasis, 88

Turkey, early agriculture in, 2, 3

turkey, domestication of, 6

turnip crop, in New Husbandry, 31, 48

Uganda, 56, 100–1; cotton in, 92; food farming in, 47, 129–30, 131; livestock in, 33, 99; mechanised Busoga farms in, 94, 129; phosphate deposits in, 96; rotation of arable and grass in, 24–5, 44

ulluca, as early food crop, 6

unemployment, in industry and agriculture, 119

United States of America, 36, 43, 86, 132; industrialisation in, 85, 116, 120; poor soils in, 89, 126

urbanisation: in Africa, 130; a drain on soil fertility, 126; in India; 108, industrialisation now possible without, 121–3; provides markets for farmers, 94, 97–8, 117

Venezuela, soils in, 44, 126

veterinary medicine, in developing countries, 32, 61, 89, 99

Vigna (cowpea), 5, 91

virus diseases of plants, 37, 42

vitamins, 62, 65, 98

volcanic soils, 23, 26, 96; rice on, 19

Warwickshire: livestock in (1935–70), 72, 73, 74, 75, 76; tillage in (1935–70), 71, 72

waste, problems of disposal of, 87, 122

water: agriculture and availability of, 14–20, 102, 134; conservation of, 9; in improvement of nomadic herding, 98; for new varieties of crop, 106, 107–8; pollution of, 86; stored in soil, 20–1, 103, 107; supply of, to conurbations, 122–3

weeds: shifting cultivation as means of controlling, 41, 45, 94; in Sudan, 43, 90

Welfare State, food prices in, 81

West Indies, 13, 31; nutrition in, 63; volcanic soils in, 23, 26

wheat, 128; climate and, 102; domestication of, 2, 5, 30; improved varieties of, 40–1, 52, 106; protein content of, 110; Yeoman variety of, 10; yields of, 40, 45, 50

wool, 31, 83

yams, 5, 91

yields: breeding crop plants for, 35, 61; of wheat, barley, and sugar beet, England (1933–69), 50; of wheat in India, 40, 107; *see also* seed/yield ratio

Zambia, shifting cultivation in, 89, 126–7

zebu, domestication of, 6